A Letter From Jack Dangermond

The collection of geographic information system (GIS) maps presented in the *ESRI® Map Book Volume 19, GIS—The Language of Geography* illustrates how GIS is contributing to almost every aspect of our lives. This includes improving business work flows, making better locational decisions, managing and conserving our resources, and providing a better scientific understanding of how our world works.

The technical capabilities of GIS have been rapidly extending to include advanced forms of spatial analysis and modeling. In addition, many of the GIS capabilities are now becoming available on the Internet and made available to everyone.

Clearly, GIS is creating new types of geographic awareness and spatial literacy. It's easy to see how GIS is evolving into a new kind of language that will cut across our scientific, professional, and education fields—a language that will provide more integrated thinking and solutions to the complex problems our society is now facing.

Warm regards,

Jack Dangermond, President, ESRI

Table of Contents

Business

4   Modeling Typhoon Risk in South Korea

5   Branch Trade Area Overlap vs. Market Financial Services Product Potential

Cartography

6   First *Space Shuttle Columbia* Disaster Media Map— Base Search Vector With Rainbow Debris Buffer

7   Base Search Vector—Regression Analysis for the *Columbia* Recovery Effort

8   *The National Map*—Prototype Graphic Product

10   Columnar Display of Multiple Attributes of a Great Salt Lake Shoreline

11   Old Fire, November 1, 2003

12   Three-Dimensional Virtual Model

13   Landing Site Maps for the Mars Exploration Rovers

14   British Columbia Topographic Series

16   Three-Dimensional Anaglyph of the Earth

Communications—Telecommunications

17   Geographic Data Technology Telecommunications and Street Data

Conservation

18   Regional Conservation Priorities for Upper Guinean and Congo Basin Forests

19   Philippine Biodiversity Conservation Priorities

20   GIS in Predicting Native American Horticulture Sites in the Allegheny National Forest

21   Monitoring Grizzly Bear Movements

22   Mapping Playa Lake Basins on the Llano Estacado, Texas

23   The Nature Audit—Cumulative Human Footprint

24   Wildfire Severity Mapping—Canberra

25   Conservation Area Hierarchy in the Meiron Mountains Reservation via GIS

26   Cultural Heritage Preservation and Zoning of Perm, Russia

28   Prioritizing Land Management in the Conservation of a Rare Species

29   Ecology and Empire Along the Ancient Silk Roads

Defense and Intelligence

30   Electrical Distribution Map, Elmendorf Air Force Base, Alaska

Education

31   ESPN SportsNation Survey

32   Quantification of Reservoir Potential of the Devonian Limestone East Waddell Ranch Survey, Texas

Environmental Management

33   Perchlorate—Drinking Water Impacts and Sources in California

34   National Oil Spill Contingency Plan for Cameroon— Vegetation Map

35   Generalized Contours of the Sauk Sequence for Characterization of Saline Aquifers for CO_2 Sequestration

36   Impact Mitigation Analysis of Helicopters on Mountain Goats in British Columbia

37   Naval Aircraft in Hampton Roads, Virginia

Government—Federal

38   Yemen Country Profile

Government—Law Enforcement and Criminal Justice

39   Sex Offender Proximity Analysis

Government—Public Safety

40   2002 Seismic Hazard Maps for the Conterminous United States

41   Earthquake Shaking Intensities for California

42   City of El Cajon Fire Department Midday Response Times

43   2003 North East/Gippsland Fires

44   City of Berkeley HAZUS Implementation

46   Estimated Snowfall and Snow Load for Colorado— Storm of March 16–20, 2003

Government—State and Local

47   Official Zoning Map, City of Roswell, Georgia

48   Yakima Traffic Maps

50   Portland Neighborhood Commercial Market Area Comparisons

51   Albury Street Map

52   Region Vysocina Administrative Structure

53   City of Toronto Building Construction Dates

54   Assessor's Map

55   Riverside County Blueprint for Tomorrow

56   Gwinnett County, Georgia, Countywide Data Integration for E911

Health and Human Services

57   West Nile Virus in Illinois—2001 and 2002

58   West Texas Atlas of Rural and Community Health

59   Aging Services Demographic Analysis

60   The Plague of Disease and Poverty in Sub-Saharan Africa

61   SARS Epidemic in China

62   Mapping Health Care Delivery for Children Using Primary Care Service Areas

Table of Contents

Natural Resources—Agriculture

63 Developing North American Soil Properties for Climate and Hydrology Applications—Mexico

64 U.S. Department of Agriculture, National Agricultural Statistics Service—Geospatial Products and Data

66 Agriculture Maps of South Africa

68 Investigating Brown Rot in Stone Fruit Using ArcIMS

Natural Resources—Forestry

69 Dinan Bay Forest Development Plan Page 1

70 Land Cover of North America

Natural Resources—Mining and Earth Science

71 A New Method to Represent Temporal Data Sets

72 Infiltration Studies

74 100+ Years of Land Change for Coastal Louisiana

75 Map of Surficial Deposits and Materials in the Eastern and Central United States (East of 102° West Longitude)

76 Seismicity Maps of the Santa Rosa Quadrangle, California, for 1969–1995

77 Sub-Andean Geological Map

78 The Application of GIS to Bauxite Mining in Jamaica

80 Czech Geological Survey—Using the National Geodatabase GEOCR50

Natural Resources— Petroleum

81 Brazil Petroleum Infrastructure

Natural Resources—Water

82 Prospect Hydroelectric Project, Rogue River, Oregon

83 California Water—21st Century Gold

84 Township of Michipicoten Municipal Groundwater Study— Quaternary Geology

85 Aeronautical Charting in New Zealand

86 Three-Dimensional Visualization of Cod Spatial Dynamics and Vertical Movements in European Waters

87 Tauranga Harbour Tidal Movements

88 Water Quality Monitoring of the Italian Rivers, a GIS Approach

89 The Watershed Fragmentation by Dams and Its Impacts on Freshwater Fishes

90 Using GIS for Data Integration and Visualization of a Deepwater Ocean Observatory

91 Managing Louisiana's Public Water Supply With GIS

92 Mapping a Diverse Population for Everglades Restoration— A Minority/Low-Income Analysis for Miami–Dade County

Sustainable Development

93 Development Through Knowledge

94 Comprehensive Planning—Town of Vinalhaven Island, Knox County, Maine

96 San Diego/Baja California Border Region View From Space

97 San Diego/Baja California Border Region Planned Land Use

Tourism

98 Colombia, Mapa Guia 2003

99 D.C. Vicinity Map and Visitor's Guide

100 Local Area Map for London Underground Stations

101 Trail of the Coeur d'Alenes

102 The Heart of Kiger Gorge, Oregon

Transportation

103 Building Tomorrow's Data Warehouse Today

104 Public Transport Optimization in the Liberec Region Project

106 Railroad Through 150 Years in Norway

107 Interstate 680 North Corridor Concept

108 Transit Commuting in America, 1990 to 2000

109 Washoe County Environmental Justice— Population and Race by Census Block

110 World Air Traffic Flowchart

Utilities—Electric and Gas

111 Electric Utility Maps

112 Kárahnjúkar Hydroelectric Project, Hálslón Reservoir

113 Electric Facilities

114 Hurricane Isabel—A Storm of Mass Destruction and an Unprecedented Restoration Effort for Dominion

115 Electric System Integration at Truckee Donner Public Utility District

Utilities—Water/Wastewater

116 Water Utility Modeling at the Truckee Donner Public Utility District

117 Preparation for a Hurricane

118 Sewer Cleaning Frequency Map

119 Sewer Mains by Date Installed

Modeling Typhoon Risk in South Korea

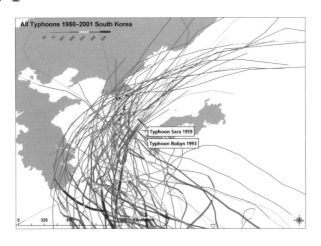

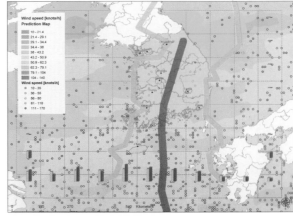

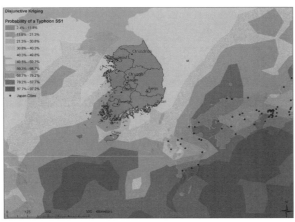

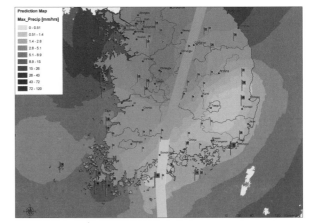

In June 2002 the Red Devils of South Korea conquered the world of football in a flash and sparked the enthusiasm of millions of fans. Barely three months later, typhoon Rusa destroyed the largest football stadium in the country. Between August 31 and September 1, 2002, South Korea was subjected to violent torrents of rain, and thousands of houses, bridges, and farmlands were destroyed. One hundred and nineteen people lost their lives to the storm.

South Korea is the seventh largest insurance market in the world. Total premium income is approximately $57 million, of which 25 percent comes from nonlife insurance companies. The insurance penetration rate is also high. Premiums as a percentage of the gross domestic product reached 10 percent for life and 3.2 percent for nonlife insurance in 2000. Statistics predict that at least one typhoon will strike the Korean peninsula each year during August.

Because of the high market penetration in South Korea, more sophisticated risk analysis is needed to determine typhoon risk. Unfortunately, the typhoon history only reveals data back to the 1950s. For an accurate risk analysis it is essential to include events with low frequencies (e.g., 100-, 250-, 1,000-year return periods) to examine the possible maximum loss and to establish a risk premium.

Typhoon data was downloaded from the Joint Typhoon Warning Center including measurements from 1950 to 2001. For each track, observations include date, time (6-hour intervals), and the main parameters of a typhoon, which are sustained wind speed, longitude and latitude, and pressure (not available for most tracks). In addition, data from 76 weather stations helped to determine the approximate wind field pattern. The wind field is used to examine the vortex of each stochastic event. To create a stochastic event set, ArcGIS® Geostatistical Analyst examined the probabilities of wind speed, forward speed, and direction of each track.

The random sampling technique determined wind speed, forward speed, and direction of each track point. All probabilities of the necessary parameters, which were determined by kriging, were defined as subgroups. Finally, parameters were selected randomly from each subgroup.

Converium Ltd.

Zurich, Switzerland

By Peter Geissbühler and Andreas Zbinden

Contact

Peter Geissbühler

peter.geissbuehler@converium.com

Software

ArcGIS Geostatistical Analyst, ArcMap™ 8.3, and Windows 2000

Printer

HP

Data Source(s)

Naval Pacific Meteorology and Oceanography Center, Joint Typhoon Warning Center

Business

Branch Trade Area Overlap vs. Market Financial Services Product Potential

Overlap vs. Product Potential

Total Potential (Millions)	1 or Less	2 to 4	5 or More
> $150			
$50 to $150			
< $50			

Overlap

Current Network Locations
- ABC Bank Offices
- XYZ Bank Offices

This map illustrates the trade area overlap resulting from the potential merger of two institutions. The block group classification categorizes the block groups by high to low financial services product potential (deposits, investments, and loans) and high to low branch trade area overlap. The resulting nine-category block group classification enables bank management to locate areas for potential consolidation or expansion.

Consolidation candidates would be located in areas with low financial services product potential and moderate to high trade area overlap. Expansion candidate areas would have low trade area overlap and moderate to high financial services product potential.

Verdi & Company
Buffalo, New York, USA
By Michael Bastedo

Contact
Michael Bastedo
michaelb@verdico.com

Software
ArcView® 8.3

Hardware
Dell Precision Workstation 360

Printer
HP Designjet 1055cm

Data Source(s)
ESRI Business Information Solutions and Verdi & Company

Business

First *Space Shuttle Columbia* Disaster Media Map—
Base Search Vector With Rainbow Debris Buffer

Forest Resources Institute, Arthur Temple College of Forestry, Stephen F. Austin State University
Nacogdoches, Texas, USA
By Jeffrey M. Williams

Contact
Jefff Williams
jmwilliams@sfasu.edu

Software
ArcGIS 8.2, ERDAS IMAGINE 8.5, and Windows XP Pro

Hardware
Dell Precision Workstation 650

Printer
HP Designjet 5000ps

Data Source(s)
Nacogdoches 911, Texas Natural Resources Information Systems, and U.S. Geological Survey

Cartography

On February 1, 2003, at approximately 8 a.m. CT the *Space Shuttle Columbia* was lost upon reentry over east Texas. Within a few minutes of the spacecraft breaking up and explosions over Nacogdoches, GIS was put to work aiding local law enforcement in protecting public safety. As debris was still raining across east Texas, geospatial scientists from the Forest Resources Institute and the Humanities Urban and Environmental Sciences (HUES) GIS Laboratories at Stephen F. Austin State University begin to map located debris with a horizontal accuracy of less than one meter using survey grade GPS units.

Within a few hours of the destruction of *Columbia,* GIS accurately modeled the shuttle's debris location and distribution by calculating a base search vector (BSV) from a least-squares linear regression using data that included Nacogdoches County 911 call sheets and "best-fit" reported debris locations. Validated by GPS data sets processed overnight by the HUES laboratory and the Center for Space Research, University of Texas, Austin, the BSV was extended across 11 counties of east Texas. Combined with detailed spatial analysis from the night of February 1 and the early hours of February 2, BSV was released to the National Aeronautics and Space Administration and local law enforcement officials during the early afternoon of February 2 in the form of a current Landsat 7 ETM+ satellite image map showing BSV surrounded by a 20-kilometer rainbow buffer of decreasing debris intensity. This was the first media map produced for the *Space Shuttle Columbia* tragedy. It was produced in mass and distributed to the media on the afternoon of February 2.

BSV was instrumental in guiding hundreds of search and recovery (SAR) teams during the critical early days of the recovery effort and resulted in the recovery of *Columbia*'s crew by February 14. Guided by BSV, the monumental SAR efforts lasted for three and one-half months involving more than 180 federal, state, and local agencies with more than 30,000 searchers covering the largest ground search area in the world—almost 700,000 acres—and recovering an unprecedented 39 percent of the shuttle's remains.

Stephen F. Austin State University geospatial analysts continued to perform rigorous analysis on the effects of subsequent debris data sets on the spatial trend of the original BSV. The analysis resulted in a drift of BSV with a trend to the south and a subtle clockwise rotation as sample size increases. In all cases, the maximum deviation from the original BSV is less than 2.3 kilometers.

The Columbia Accident Investigation Board's final report recognizes the contributions of Stephen F. Austin State University and the critical role that GIS played in the early days of the recovery efforts . Although reconstructed many times, the BSV developed from vastly limited data during the first crucial days of the *Columbia* tragedy proved to be an accurate guide that allowed the historic SAR effort to recover enough of the shuttle's remains for the *Columbia* Accident Investigation Board to identify the probable cause of the mission's reentry failure.

Base Search Vector—Regression Analysis for the *Columbia* Recovery Effort

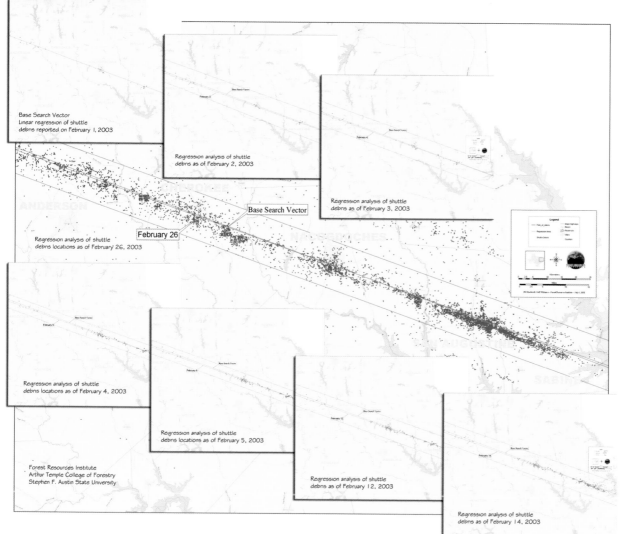

Forest Resources Institute, Arthur Temple College of Forestry, Stephen F. Austin State University

Nacogdoches, Texas, USA

By P.R. Blackwell

Contact

P.R. Blackwell

prblackwell@sfasu.edu

Software

ArcGIS 8.2, ERDAS IMAGINE 8.5, and Windows XP Pro

Hardware

Dell Precision Workstation 650

Printer

HP Designjet 5000ps

Data Source(s)

Nacogdoches County 911, Texas Natural Resources Information Systems, and U.S. Geological Survey

---

Cartography

The National Map—Prototype Graphic Product

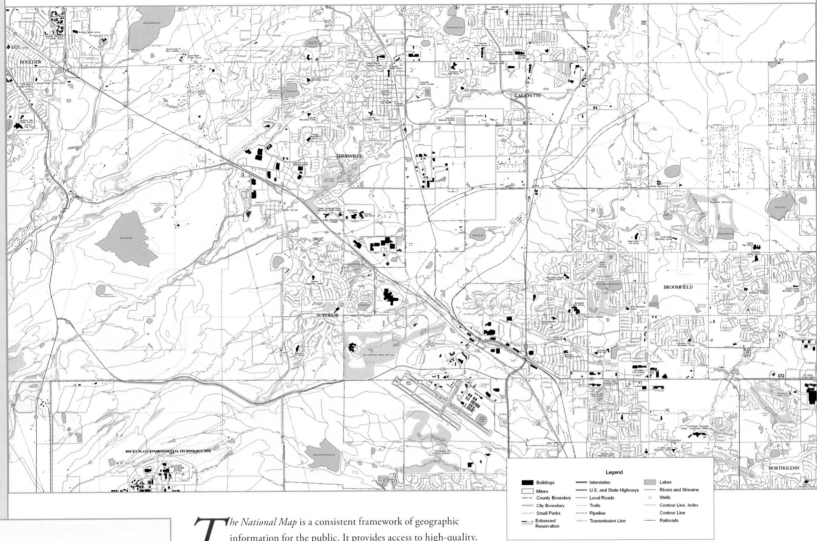

U.S. Geological Survey, Rocky Mountain Mapping Center

Denver, Colorado, USA

By Mark Bauer, Alexis Ellis, and Jennifer Stefanaci, USGS and Stanley Wilds, Parallel Inc.

Contact

Stan Wilds

srwilds@usgs.gov

Software

ArcMap, ArcSDE, Maplex, and ERDAS IMAGINE

Printer

HP Designjet 5000

Data Source(s)

Colorado Department of Public Health, Colorado Department of Transportation, Colorado Office of Emergency Management, Federal Emergency Management Agency, U.S. Department of Transportation, and USGS

Cartography

*T*he *National Map* is a consistent framework of geographic information for the public. It provides access to high-quality, geospatial data and information from multiple partners to help support decision making. The vision is that, by working with partners, the public will have access to current, accurate, and nationally consistent digital data and topographic maps.

The 1:24,000-scale topographic map has been the primary U.S. Geological Survey (USGS) map series product for nearly a century. Historically, this map was created manually from newly derived spatial data. USGS has been developing new topographic-like map products using existing current digital data collected and integrated through partnerships with state and local agencies. The creation of prototype graphic products assumes that new products can be produced, maintained, and updated in a reduced amount of time at lower costs than by traditional means. Further, this product is designed to be generated as geocoded raster products as well as hard-copy maps.

The National Map prototype 1:24,000-scale maps were produced from integrated partnerships and USGS data covering the Denver metropolitan and surrounding areas. The data, all 1:24,000 scale or better, represents local, state, and federal cooperative data sets. Data was collected and integrated as part of the USGS 133-city Denver pilot project and is stored in ArcSDE® at the USGS Rocky Mountain Mapping Center. The maps depict the USGS Lafayette/Louisville, Colorado, and Conifer, Colorado, 7.5-minute series areas and illustrate the seamless quality and other properties of the data.

These prototype graphics incorporated ArcMap tools for color generation, uniform symbology, and raster graphic generation. To produce advanced, flexible text placement without conflicts, Maplex tools were used in the production process.

The National Map—Prototype Graphic Product

U.S. Geological Survey, Rocky Mountain Mapping Center

Denver, Colorado, USA

By Mark Bauer, Alexis Ellis, and Jennifer Stefanaci, USGS and Stanley Wilds, Parallel Inc.

Contact

Stan Wilds

srwilds@usgs.gov

Software

ArcMap, ArcSDE, Maplex, and ERDAS IMAGINE

Printer

HP Designjet 5000

Data Source(s)

Colorado Department of Public Health, Colorado Department of Transportation, Colorado Office of Emergency Management, Federal Emergency Management Agency, U.S. Department of Transportation, and USGS

Cartography

Columnar Display of Multiple Attributes of a Great Salt Lake Shoreline

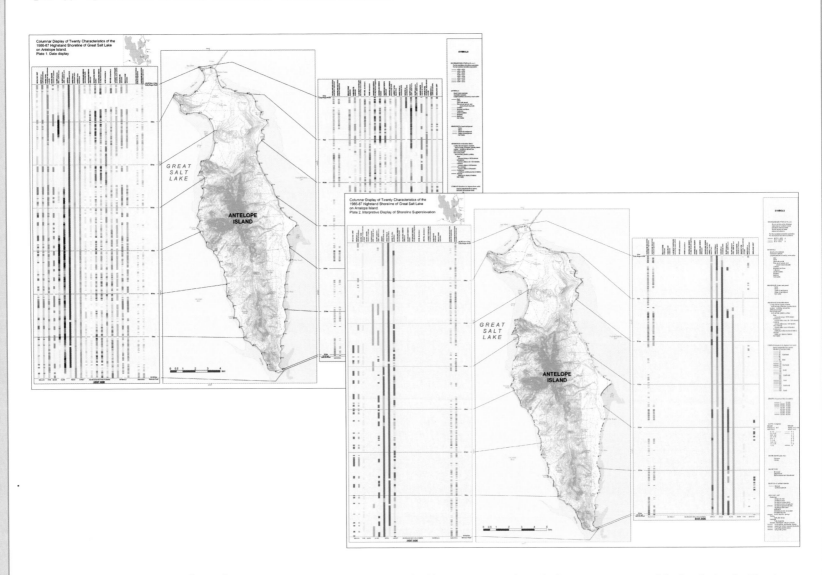

University of Utah

Salt Lake City, Utah, USA

By Genevieve Atwood

Contact

Genevieve Atwood

genevieve.atwood@geog.utah.edu

Software

ArcView® 8.2, and Windows XP

Hardware

Gateway 2000

Printer

HP Designjet 800ps

Data Source(s)

Dissertation research in progress

Cartography

These images show one approach to a challenge faced by many researchers—how to visualize abundant data that describes diverse characteristics of a complex, linear feature. Multiple columns display multiple attributes of a shoreline on Antelope Island, the largest island in Great Salt Lake. Location is referenced in two spatial systems—geographic coordinates and distance along a linear-referenced, 64-kilometer shore route measured clockwise from the northern tip of the island. The shore route with kilometer annotation is displayed on the map of Antelope Island. Grid lines associate location along the shore route with position on a column. Because Antelope Island is elongated north and south, columns are displayed for the two sides of the island.

Plate 1 displays 20 characteristics of the shoreline created during 1986 and 1987 by the flooding of Great Salt Lake. Each column displays one shoreline characteristic. Shoreline characteristics such as elevation, length of maximum fetch, and shore zone slope are quantitative data. Characteristics such as relative abundance of shoreline debris are ordinal data. Nominal data includes shore shape and shore processes. For each column, color indicates attribute values of the shoreline characteristic.

Plate 2 displays one interpretation of the data displayed in Plate 1. In this example, relationships of 18 of the shoreline characteristics of Plate 1 have been analyzed visually for their association with the variable, shoreline superelevation. Shoreline superelevation is the elevation difference between a shoreline created by a water body and the elevation of the water body's still water surface undisturbed by wind, waves, and currents. Superelevation of the 1986–1987 shoreline of Antelope Island varies approximately four meters around the island. Plate 2 is one snapshot of an analysis of one variable of the Antelope Island data set. In ArcMap each column is a map layer that a researcher can interactively display, hide, reorder, and color to explore relationships among variables.

Columnar display of multiple attributes is a GIS technique not limited to shore zones and coastal processes. The technique can be used to visualize complex and diverse data associated with linear features such as fault zones, core logs, political boundaries, highways, and streams. For further explanation of the technique, see http://gis.esri.com/library/userconf/proc03/p0974.pdf.

Old Fire, November 1, 2003

In October 2003, fires raged throughout Southern California in San Diego, San Bernardino, Los Angeles, Ventura, and Riverside counties. They scorched hundreds of thousands of acres, displaced tens of thousands of people, and destroyed thousands of homes. As emergency responders worked to put out the fires with every resource available, computer mapping played a key role in many areas.

This map was created by ESRI's technical marketing team for the San Bernardino County Sheriff's department to assist the sheriff's emergency operations center. The map shows the current Old Fire and Grand Prix Fire perimeters on November 1, 2003. The fire perimeter data was captured using aerial equipment and GPS. The perimeter is overlaid with digital elevation data and local transportation data.

ESRI
Redlands, California, USA
By Mark Reddick

Contact
Tim Rankin
trankin@esri.com

Software
ArcGIS 9

Data Source(s)
Aerial photography and others

Cartography

Three-Dimensional Virtual Model

An Example of a Community Model

Creation of a Complex Object: A Church

Creation of a Digital Terrain Model

An Orthophoto of a Digital Terrain Model   An Orthophoto of a Digital Surface Model

Creation of a Digital Surface Model

Construction

This map depicts a digital terrain model and a digital elevation model with buildings added manually. Many research threads were not presented in this poster because they are in the test phase. Urban analysis in two dimensions reaches its limits of interpretation, but analysis can be expanded when the Z dimension is integrated. A set of constraints on space, such as noisy roads, highways, and airport zones, can become apparent when it materializes by volume.

Representation in 3D and Overlay

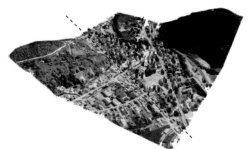

A 3D Representation

ADAUHR

Colmar, France
By Dominique Esnault

Contact

Dominique Esnault
desnault@adauhr.asso.fr

Software

ArcGIS 3D Analyst™ and
ArcGIS Spatial Analyst

Hardware

HP

Printer

HP Designjet 5500

Data Source(s)

ADAUHR and IGN © BD

Cartography

Landing Site Maps for the Mars Exploration Rovers

With the successful landing of the two Mars exploration rovers, *Spirit* and *Opportunity*, Mars has once again commanded the attention of cartographers around the world. As of May 2004, both rovers have met their 100 percent "mission success" goals. With both rovers working at full capacity, the project was extended into September 2004. These two figures display the landing site locations of the two rovers using remotely sensed data. Two National Aeronautics and Space Administration (NASA) satellites circling Mars, *Mars Global Surveyor* and *Mars Odyssey*, gathered the images and topographic data used.

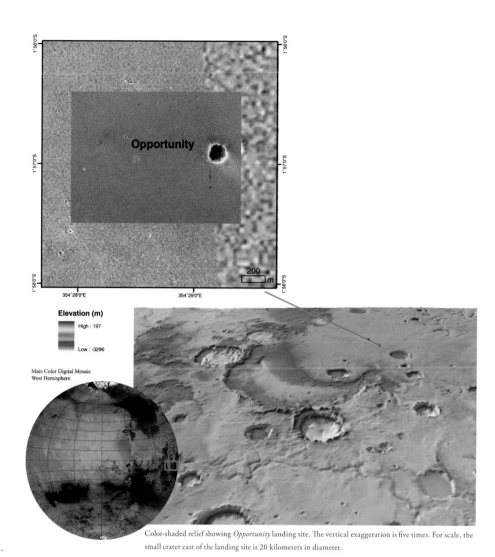

Color-shaded relief showing *Opportunity* landing site. The vertical exaggeration is five times. For scale, the small crater east of the landing site is 20 kilometers in diameter.

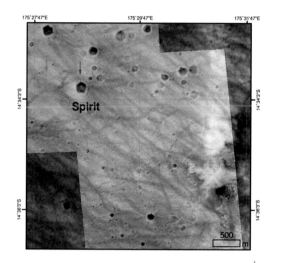

Color-shaded relief showing Gusev crater and the large channel Ma'adim Vallis flowing into it from the south. The vertical exaggeration is five times. For scale, Gusev crater is approximately 150 kilometers wide from rim to rim.

U.S. Geological Survey, Astrogeology Team
Flagstaff, Arizona, USA
By Trent Hare

Contact
Trent Hare
thare@usgs.gov

Software
ArcMap, Adobe Illustrator, and Integrated Software for Imagers and Spectrometers

Hardware
PC

Data Source(s)
Arizona State University, Cornell University, Jet Propulsion Laboratory, Malin Space Science Systems, NASA, and U.S. Geological Survey

Cartography

British Columbia Topographic Series

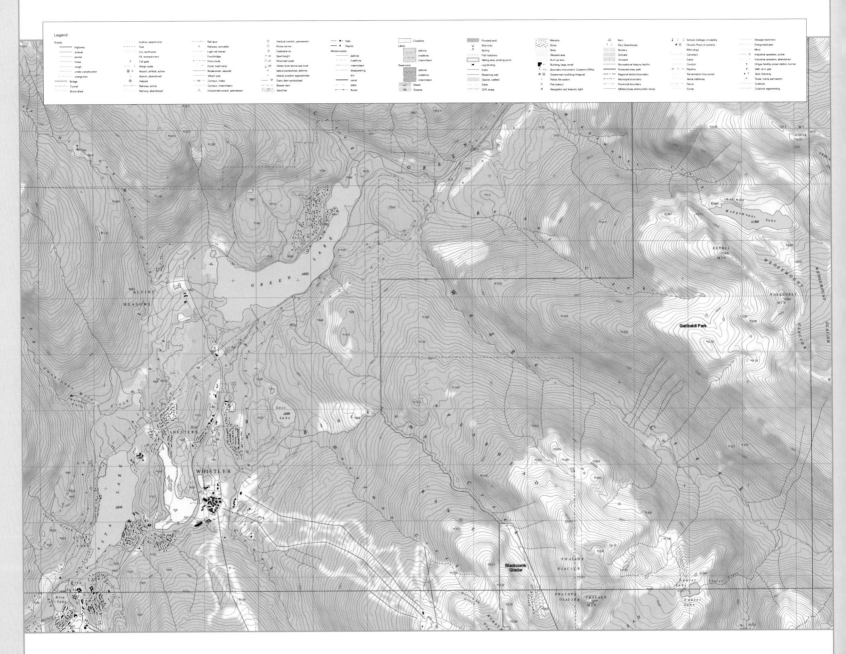

Clover Point Cartographics Ltd.

Victoria, British Columbia, Canada

By Kathleen Lush and Mike Shasko

Contact

Mike Shasko

mike.s@cloverpoint.com

Software

ArcInfo® 8.1 and Windows 2000

Printer

HP Designjet 1055cm

Data Source(s)

British Columbia Terrain Resource
Information Mapping and in-house data

Cartography

British Columbia Topographic Map Series

The British Columbia Topographic maps are a new 1:20,000 series covering the province's 7,016 maps. Clover Point Cartographics Ltd. has developed an enhanced cartographic product detailing terrain and resource databases.

As a commercial product, much of the effort was directed at developing a consistent and intuitive geographic data look for both the general and advanced user. As much of British Columbia is relatively remote, these maps provide an exciting view of the landscape for avid adventurers, resource planners, and all in between.

British Columbia Topographic Series

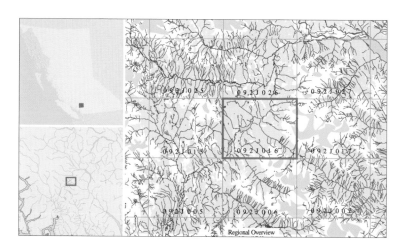

Regional Overview

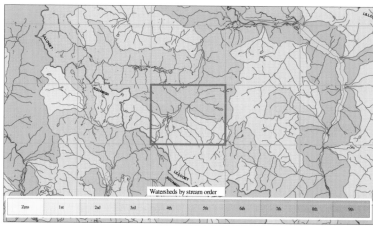

Watersheds by stream order

Zero	1st	2nd	3rd	4th	5th	6th	7th	8th	9th

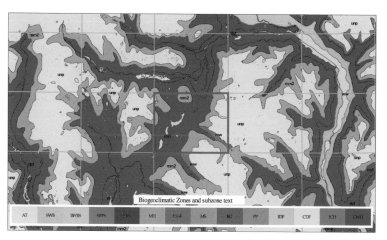

Biogeoclimatic Zones and subzone text

AT	SWB	BWBS	SBPS	SBS	MH	ESSF	MS	BG	PP	IDF	CDF	ICH	CWH

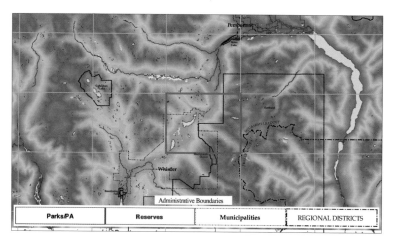

Administrative Boundaries

Parks/PA	Reserves	Municipalities	REGIONAL DISTRICTS

Clover Point Cartographics Ltd.
Victoria, British Columbia, Canada
By Kathleen Lush and Mike Shasko

Contact
Mike Shasko
mike.s@cloverpoint.com

Software
ArcInfo 8.1 and Windows 2000

Printer
HP Designjet 1055cm

Data Source(s)
British Columbia Terrain Resource
Information Mapping and in-house data

Cartography

Three-Dimensional Anaglyph of the Earth

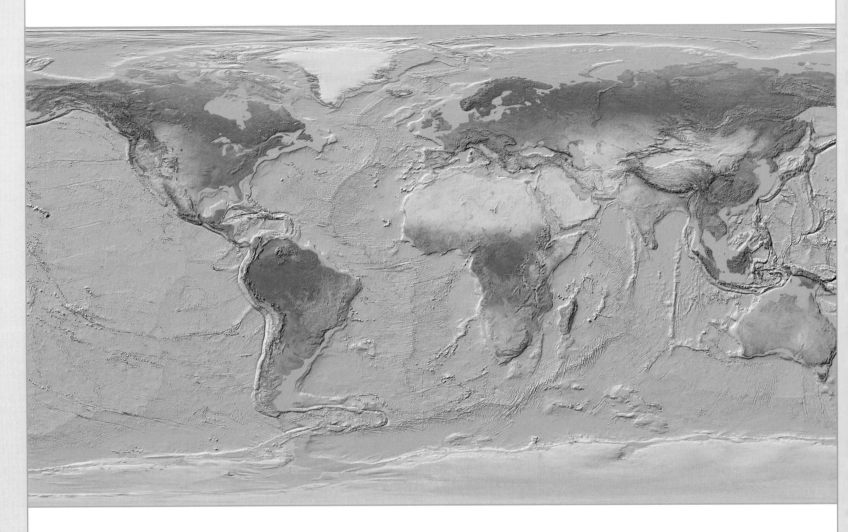

U.S. Geological Survey

Sioux Falls, South Dakota, USA
By Brian Davis, Paul Morin, and
Monte Ramstad

Contact

Brian Davis
bdavis@usgs.gov

Printer

Light Jet

Data Source(s)

NASA, Smith/Sandwell, and USGS

Cartography

At first glance, the map in front of you may seem rather ordinary. But, put on a pair of 3D glasses and see what happens. Mountains rise up and valleys recede, as this two-dimensional map appears to become three-dimensional. This map is the product of collaboration between the U.S. Geological Survey (USGS) Earth Resources Observation Systems (EROS) Data Center staff and researchers at several other institutions including the University of Minnesota. These scientists are creating innovative maps—both printed and digital—that make it possible to view Earth's geographic features in unique ways. These visualizations become valuable tools for understanding complex geospatial relationships.

The data about land surfaces and coastal regions that was necessary to make this map was gathered by a satellite sensor called MODIS (short for moderate resolution imaging spectroradiometer). MODIS is carried aboard the National Aeronautics and Space Administration (NASA) Terra satellite, which is currently orbiting approximately 700 kilometers (435 miles) above the Earth's surface. Techniques developed by USGS combine the MODIS satellite data with elevations of the land surface and seafloor. The result is this "global mosaic" that reveals the topographic features of our planet's surface in remarkable detail. Seafloor data was derived from various declassified U.S. Navy sources. These sources include satellite altimetry observations combined with shipboard echo sounding measurements to ensure accuracy.

How is the 3D effect achieved? Using the data, two separate views of the Earth's surface are generated, each at a slightly different perspective. These two views are color coded in red and blue and superimposed onto each other to make a type of picture called an anaglyph. It produces a three-dimensional effect when viewed through the corresponding red-and-blue colored filters of the special glasses.

Geographic Data Technology
Telecommunications and Street Data

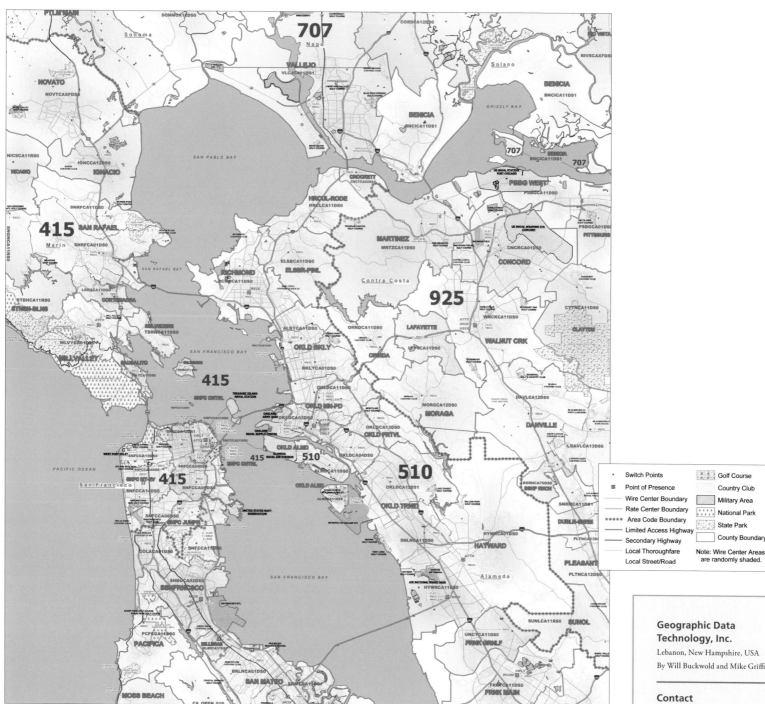

L ocal Exchange Carrier (LEC) boundaries enable users to determine market penetration by viewing the boundary extent of competitive telecommunication corporations and enable customers to analyze existing coverage of telephone service, compare markets, and examine market penetration. LECs represent a service area for a telecommunication utility that has been granted either a certificate of convenience, necessity, or operating authority to provide local exchange telephone service.

Exchange boundaries define the rate center area. This is the area that a company will charge a certain rate for service. The exchange area is synonymous with the rate center.

Wire centers (WC) represent a location, usually a building, containing telephone switching equipment. The WC boundary is a representation of the area served by the location. WC points are also provided in this product and represent the actual location of the primary central office.

Geographic Data Technology, Inc.

Lebanon, New Hampshire, USA

By Will Buckwold and Mike Griffin

Contact

Deb McCaffrey

deborah_mccaffrey@gdt1.com

Software

ArcMap 8.3 and Windows NT

Printer

HP Designjet 1055cm

Data Source(s)

Geographic Data Technology, Inc.

Communications— Telecommunications

Regional Conservation Priorities for Upper Guinean and Congo Basin Forests

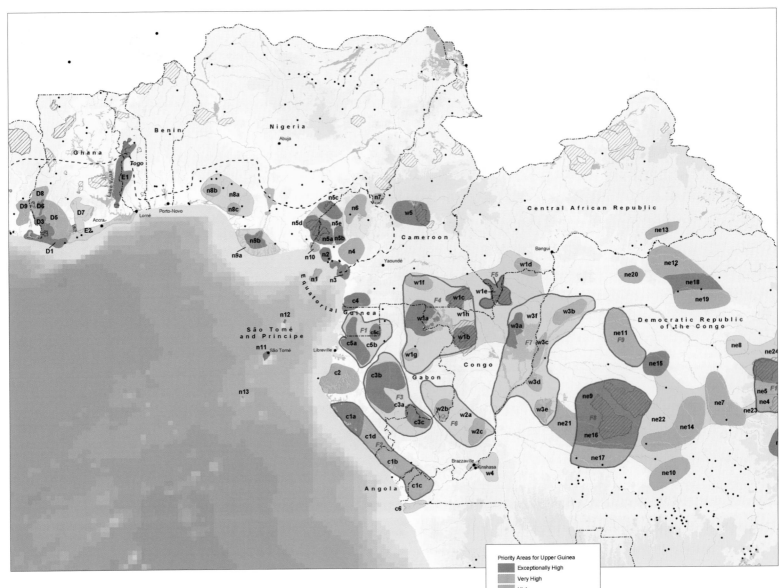

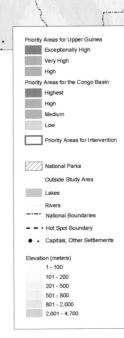

Conservation International
Washington, D.C., USA
By Sonya Krogh

Contact
Mark Denil
m.denil@conservation.org

Software
ArcGIS and Windows 2000

Printer
HP Designjet 5000ps

Data Source(s)
ESRI, Conservation International, United
Nations Environment Programme, and
World Conservation Monitoring Centre

Conservation

**Center for Applied Biodiversity Science at
Conservation International**

This map depicts the regional conservation priorities from the separate workshops organized by Conservation International for the Upper Guinea forest ecosystem and World Wildlife Fund for the Congo Basin forest.

The Upper Guinea workshop included Guinea, Sierra Leone, Liberia, Ghana, Cote d'Ivoire, and Togo. The Congo Basin workshop extended into Nigeria where priority sites were also identified.

Priority Areas for Upper Guinea
- Exceptionally High
- Very High
- High

Priority Areas for the Congo Basin
- Highest
- High
- Medium
- Low

- Priority Areas for Intervention
- National Parks
- Outside Study Area
- Lakes
- Rivers
- National Boundaries
- Hot Spot Boundary
- Capitals, Other Settlements

Elevation (meters)
- 1 - 100
- 101 - 200
- 201 - 500
- 501 - 800
- 801 - 2,000
- 2,001 - 4,700

Philippine Biodiversity Conservation Priorities

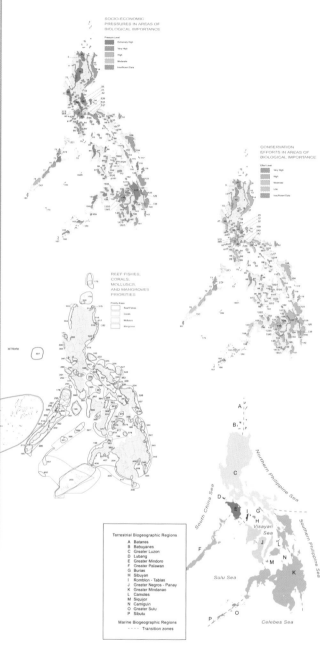

Conservation International

Washington, D.C., USA

By Philippine Biodiversity Conservation
Priority-setting Program

Contact

Mark Denil

m.denil@conservation.org

Oliver G. Coroza

ocoroza@conservation.org

Software

ArcView 3.2, Adobe Illustrator, Avenza
MAPublisher, and Windows NT and 2000

Hardware

PC desktop workstations

Printer

Offset press

Data Source(s)

Digital Chart of the World, National
Mapping and Resource Information Authority
(Philippines), Philippine Biodiversity
Conservation Priority-setting Program,
and other national and local Philippine
organizations and international institutions

Conservation

The Philippines is in the midst of a biodiversity crisis. It is among the world's richest countries in biodiversity, but its biodiversity is threatened. The results shown here indicate the areas that must be conserved to thwart further loss of biodiversity.

One of the major contributions of the Philippine Biodiversity Conservation Priority-setting Program was the updating of the country's biogeographic regions. Sixteen terrestrial and six marine biogeographic regions were identified. This consensus was based on recent information on the Philippine's geologic and evolutionary history and experts' opinion about the distribution patterns of various taxonomic groups. This effectively updated biogeographic zones previously identified by the Department of Environment and Natural Resources in 1997.

Each of the 16 terrestrial biogeographic regions, which is a separate island or island group that existed during the last Ice Age, supports a large number of unique species and is recognized as a center of biodiversity. The present Philippine Islands arcs surrounding the important basins in the South China Sea, the Sulu Sea, and the Pacific Ocean served as the physical framework for the marine biogeographic regions. The six marine biogeographic regions were identified with broad transition zones based on the affinities of the associated reef fish assemblages, the evolutionary geology of the archipelago, and the predominant ocean circulation patterns.

GIS in Predicting Native American Horticulture Sites in the Allegheny National Forest

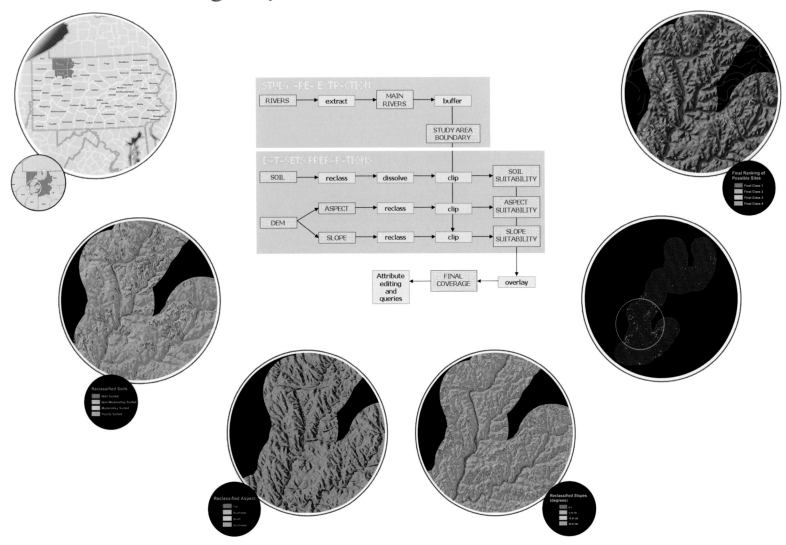

Clarion University

Clarion, Pennsylvania, USA

By Andrew Angel and Yasser Ayad

Contact

Yasser Ayad

yayad@clarion.edu

Software

ArcInfo 8.3, Adobe Photoshop, Microsoft Excel, Microsoft PowerPoint, and Windows XP

Hardware

Pentium PC

Printer

HP Designjet 1055cm

Data Source(s)

Pennsylvania Spatial Data Access and U.S. Geological Survey

Conservation

The Allegheny River Basin is widely known for containing many Native American occupation sites along the river's banks and valleys throughout western Pennsylvania. The river was a main travel route, acting like a major interstate, for various bands of tribes that inhabited northwestern Pennsylvania and the Finger Lakes area of New York.

This poster presents a project that involved the use of multiple data sets to represent site characterizations for Native American horticulture, specifically corn production. The location and characteristics of such sites can be identified by specific conditions such as soil suitable for agriculture and landform properties such as slope and orientation of land surface (i.e., aspect). Pinpointing those sites could lead to the discovery of occupation zones such as villages and/or smaller residential agglomerations, which usually are located in proximity to the agricultural fields. Potential rock shelters were also included in the analysis of the potential locations for future archaeological excavation.

Data was collected and compiled for areas within four miles from the Allegheny River and reclassified to represent soils suitable for corn production, relatively flat areas, and regions that face south. Raster modeling was carried out to overlay the reclassified soil, slope, and aspect maps, and attributes were queried for different combinations of site conditions. Many regions were identified as potential corn production sites, and possible rock shelters were extracted from a surface geology map. They were buffered and intersected with the previously identified sites, and the resulting polygons were recognized to be the most probable locations for future archaeological excavation.

Although further field tests were not carried out to confirm the results, the findings of the current study coincided with existing archaeological sites in the West Hickory region, located south of the study area. This study demonstrates that GIS modeling is a promising tool for archaeologists especially if it is strengthened with field investigations and integrated with the results of site characterization of existing excavations.

Monitoring Grizzly Bear Movements

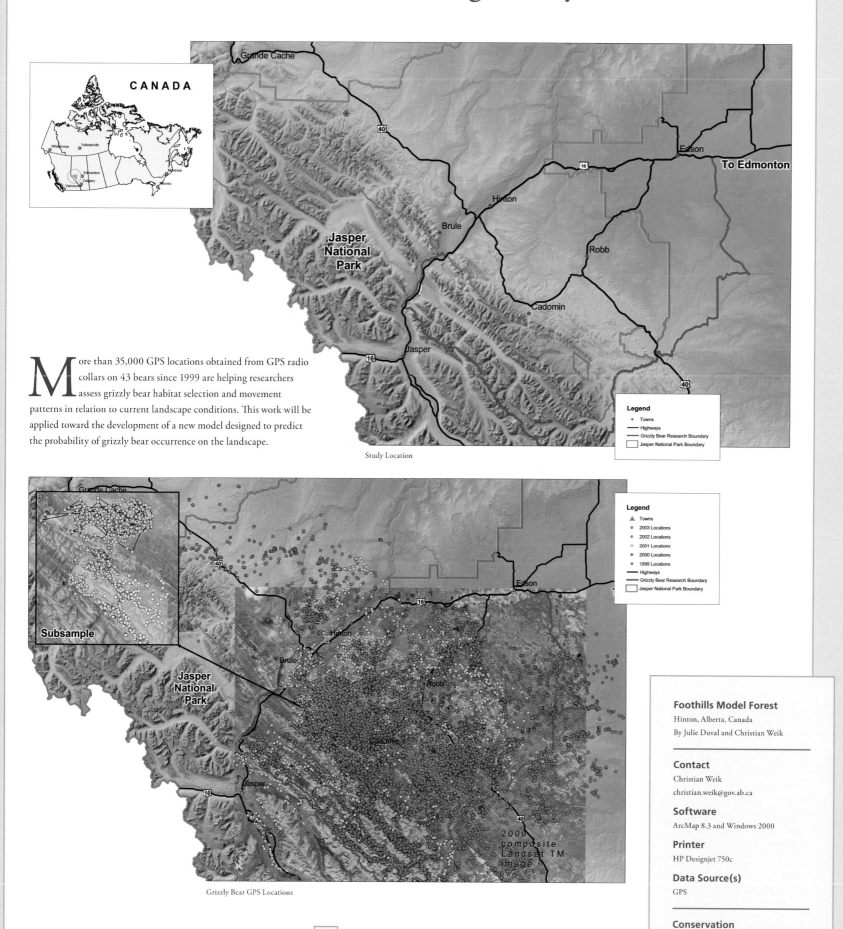

More than 35,000 GPS locations obtained from GPS radio collars on 43 bears since 1999 are helping researchers assess grizzly bear habitat selection and movement patterns in relation to current landscape conditions. This work will be applied toward the development of a new model designed to predict the probability of grizzly bear occurrence on the landscape.

CANADA

Study Location

Legend
- Towns
- Highways
- Grizzly Bear Research Boundary
- Jasper National Park Boundary

Legend
- Towns
- 2003 Locations
- 2002 Locations
- 2001 Locations
- 2000 Locations
- 1999 Locations
- Highways
- Grizzly Bear Research Boundary
- Jasper National Park Boundary

Subsample

2000 composite Landsat TM image

Grizzly Bear GPS Locations

Foothills Model Forest
Hinton, Alberta, Canada
By Julie Duval and Christian Weik

Contact
Christian Weik
christian.weik@gov.ab.ca

Software
ArcMap 8.3 and Windows 2000

Printer
HP Designjet 750c

Data Source(s)
GPS

Conservation

Mapping Playa Lake Basins on the Llano Estacado, Texas

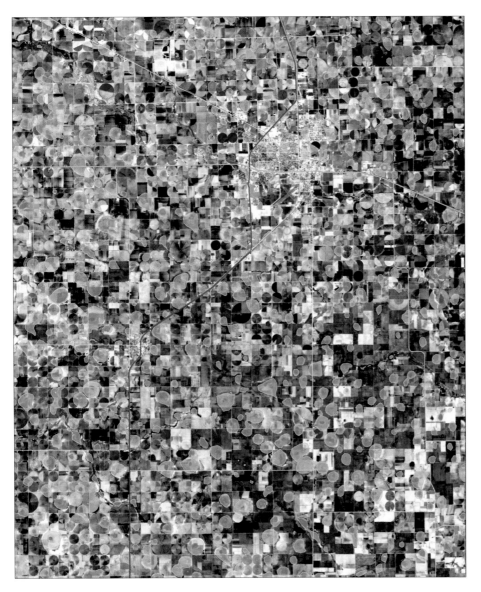

DIGITAL ELEVATION MODEL FOR HALE COUNTY

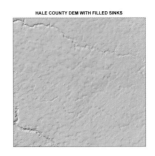

HALE COUNTY DEM WITH FILLED SINKS

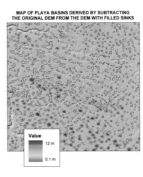

MAP OF PLAYA BASINS DERIVED BY SUBTRACTING
THE ORIGINAL DEM FROM THE DEM WITH FILLED SINKS

Value
12 m
0.1 m

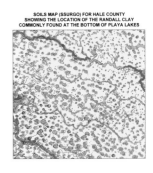

SOILS MAP (SSURGO) FOR HALE COUNTY
SHOWING THE LOCATION OF THE RANDALL CLAY
COMMONLY FOUND AT THE BOTTOM OF PLAYA LAKES

THE DISTRIBUTION OF RANDALL CLAY
EXTRACTED FROM THE SSURGO SOILS DATA

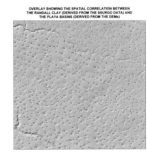

OVERLAY SHOWING THE SPATIAL CORRELATION BETWEEN
THE RANDALL CLAY (DERIVED FROM THE SSURGO DATA) AND
THE PLAYA BASINS (DERIVED FROM THE DEMs)

DISTRIBUTION OF PLAYA LAKES IN WEST TEXAS
BASED UPON THE OCCURRENCE OF RANDALL CLAY

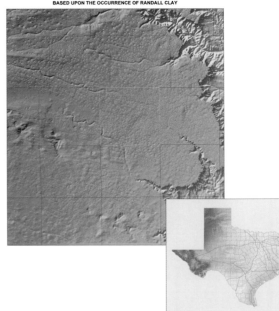

Texas Tech University

Lubbock, Texas, USA

By Ernest Fish and Kevin Mulligan

Contact

Kevin Mulligan

kevin.mulligan@ttu.edu

Software

ArcInfo 8.3

Hardware

Dell workstation

Printer

HP Designjet 800ps

Data Source(s)

ESRI, National Elevation Dataset, and
internal sources

Conservation

The Llano Estacado is a relatively flat tableland located on the southern High Plains in west Texas. The region is unique insofar as the landscape is characterized by thousands of ephemeral playa lakes and few streams.

Understanding the spatial distribution and size of playa basins is extremely important in this semiarid landscape. The playas provide a water source and critical habitat for local wildlife and migratory birds, and the thousands of playa lakes serve as groundwater recharge basins for the southern Ogallala aquifer.

To map the distribution of playa basins, the Fill Sinks function of Arc Hydro was used to fill the depressions in the digital elevation model (DEM) for Hale County. The original DEM was then subtracted from the new DEM with the filled sinks. Where no depressions exist, the grid cells subtract to zero. Where depressions occur, the grid cells map the extent, depth, and volume of the playa basin.

The Nature Audit—Cumulative Human Footprint

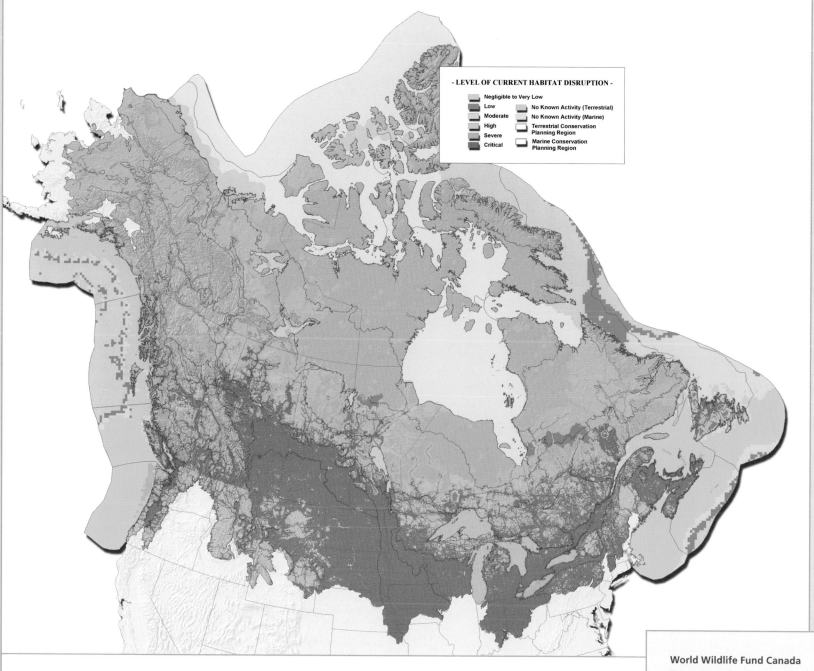

- LEVEL OF CURRENT HABITAT DISRUPTION -

Negligible to Very Low

Low

Moderate

High

Severe

Critical

No Known Activity (Terrestrial)

No Known Activity (Marine)

Terrestrial Conservation Planning Region

Marine Conservation Planning Region

Released by World Wildlife Fund Canada in May 2003, the Nature Audit is a groundbreaking report that audits Canada's efforts to conserve biodiversity and provide an action agenda for the 21st century. Though the report found that opportunities to protect large ecosystems in more than 50 percent of Canada have already been lost, the Nature Audit highlighted strategies to protect, manage, and restore the nation's natural capital.

As a part of the report, this map shows the current cumulative impact or footprint of human activities on wildlife through disruption to their habitat and ecosystem integrity in Canada's ecoregions. For more information, visit www.wwf.ca.

World Wildlife Fund Canada
Toronto, Ontario, Canada
By Alexis Morgan

Contact
Alexis Morgan
amorgan@wwfcanada.org

Software
ArcMap

Printer
HP 2500c

Data Source(s)
ESRI, U.S. Geological Survey, World Wildlife Fund, and other sources

Conservation

Wildfire Severity Mapping—Canberra

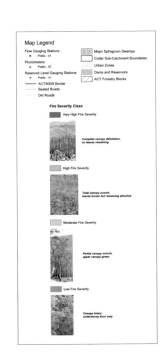

ECOWISE Environmental
Fyshwick, Canberra, Australia
By ECOWISE Fire Recovery Team

Contact
Tanya Whiteway
twhiteway@ecowise.com.au

Software
ArcGIS 8.2, ENVI 3.6 RT, and
Windows 2000

Hardware
Dual Athlon 2000

Printer
HP Designjet 1055cm

Data Source(s)
Australian Surveying and Land
Information Group, ECOWISE
Environmental, Landsat Enhanced
Thematic Mapper+, New South Wales Land
and Property Information, and New South
Wales National Parks and Wildlife Service.

Conservation

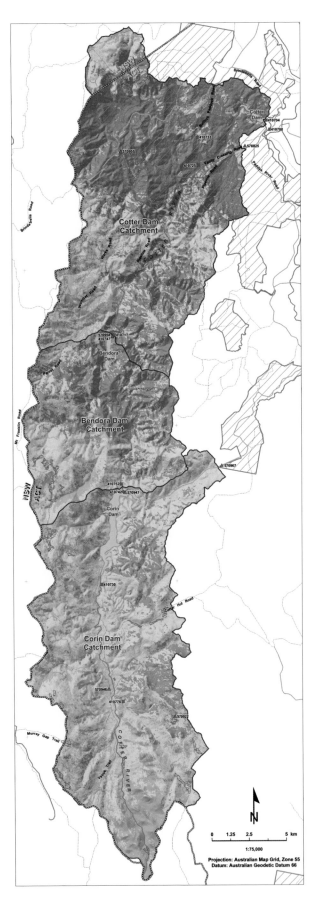

In January 2003, fires spread throughout the eastern states of Australia causing evacuations, loss of livestock, and loss of property. In the Australian Capital Territory (ACT), fires were particularly devastating because of the rate of spread and the area they covered. On January 8, electrical storms ignited five fires in the Brindabella range to the east of Canberra. All five fires remained small, burning approximately 70 square kilometers (27 square miles) in the eight days following their ignition. On January 18, fires fanned by winds reaching speeds of 100 kilometers per hour, and hot, dry conditions burned 1,649 square kilometers. Some 80 percent of ACT were fire affected.

The wildfires resulted in the loss of four lives, 500 homes, and 50,000 electricity and 7,000 gas customers temporarily lost their services. In addition, 100 percent (200 square kilometers) of the primary water catchment (Cotter Catchment) was burned. Key water treatment processes were inoperable for nine hours, and sewage treatment processes were inoperable for two and one-half days.

In the years prior to the 2003 fires, the Cotter Catchment supplied up to 95 percent of water for more than 350,000 Canberra residents. The catchment formed an essential barrier to water pollutants, both naturally occurring and anthropogenic, and the ecosystem services it provided made the ACT water supply one of the most pure in Australia, requiring only lime pH adjustment, chlorination, and fluoridation.

In many of the most hydrologically significant areas, the catchment had been completely denuded of vegetation, both the understory and the canopy. The loss of vegetation and swamps during the fires has dramatically changed the water quality at the reservoir. Turbidity, iron, and manganese are now water quality issues. In response to the degradation of water quality, a new treatment plant will be constructed. In keeping with the multibarrier approach to maintaining water quality, time and resources have been focused on the catchment to assist with rehabilitation and future management.

The inaccessible nature of the catchment meant that remote sensing was an ideal tool for initial rapid assessment. The methodologies employed in the fire severity mapping were developed toward an industry standard with other authorities and organizations. The purpose of this analysis was to provide catchment managers with a method of prioritizing catchment activities to manage the effects of fires on water quality and hydrology. Prefire and postfire Landsat Thematic Mapper images were analyzed to create a data set showing the severity of the burn in the catchment.

Remote sensing analysis will be repeated in 2004 to monitor vegetation recovery.

Conservation Area Hierarchy in the Meiron Mountains Reservation via GIS

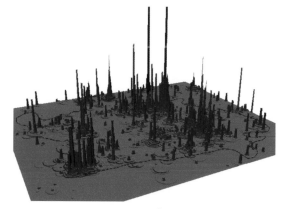

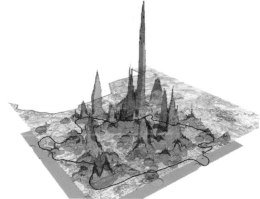

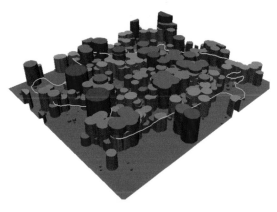

The purpose of this work was to design a GIS as a support tool for environmental and planning conflicts. At the core of this system lie the evaluation and analysis of the natural and landscape resources of the Har Meiron reservation while emphasizing the reservation's importance and its role in preserving valuable natural resources.

The Har Meiron reservation is the focal point of a conflict for the last 17 years. The conflict stems from citizens who live in villages that are positioned at the middle of the natural reserve. They want to further develop their lands in the reserve, but the Ministry of the Environment and other "green" organizations, such as the Society for the Protection of Nature and the Reservation Authority, reject their plans preferring to keep the reservation as is.

The spatial database for the research was built from different data coverages collected from several sources. The data was classified and structured into a few major layers such as flora, built areas, roads, archaeology sites, and landmarks.

At the end of the process, all layers were incorporated into one grid layer that presented area cells with the following order of conservation levels: (1) area cells with the highest conservation level; (2) area cells with intermediate conservation level; (3) area cells with the lowest conservation level; (4) already developed areas; (5) the conflict areas—developed area cells that were ranked as high preservation level; and (6) areas that are undeveloped and do not contain natural values. In the future, it will be possible to create other spatial analyses to support decision making for regional sustainable development planning, and this system may serve as a solution for other environmental conflicts at the planning committee phase.

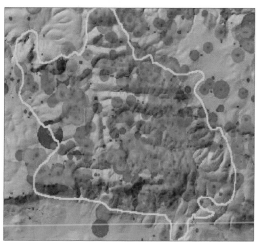

Haifa University, GIS and Remote Sensing Lab
Haifa, Israel
By Ammatzia Peled and Yoav Steinberg

Contact
Yoav Steinberg
ulz25@yahoo.com

Software
ArcInfo 7.1, ArcView 3.2a, ArcView 3D Analyst, ArcView Spatial Analyst, and Windows 98 SE

Printer
HP Designjet 750c

Data Source(s)
Israel National GIS and Rotem–Israel Plants Information Center

Conservation

Cultural Heritage Preservation and Zoning of Perm, Russia

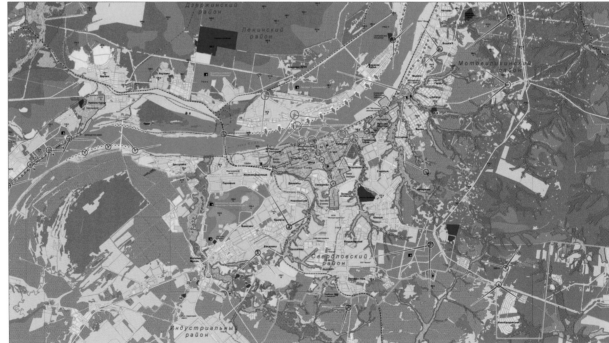

Scientific and Projecting Institute of Spatial Planning ENKO

St. Petersburg, Russia
By the ENKO stafff

Contact

Olga Krasovskaya
olgakrasovskaya@mail.ru

Software

ArcView 3.2, ArcView 8.2, Avenue™, Linux RH, and Oracle9*i* database for Linux/IA64

Hardware

Intel PIV 2.0

Printer

HP Designjet 120 and HP Deskjet 1220

Data Source(s)

Department of Architecture and Planning

Conservation

Perm is an administrative and industrial center of approximately 800 square kilometers in the northeast European part of Russia, located in the foothills of the Ural Mountains on the Kama River. It was founded in 1723 and currently has a population of approximately one million. There are more than 400 historical and cultural places including 30 archaeological sites.

GIS is used for the development of town planning projects such as the master plan, cultural heritage preservation, and zoning. A multipurpose database of Perm town planning and historical and cultural objects was created for various projects. A multifunctional analysis of the city landscape was undertaken with the use of GIS and remote sensing imagery. Separate thematic GIS layers, such as historical zoning, the stages of Perm town planning development, and culturally significant sites, are shown on the maps. The city also uses GIS software for territorial development management.

Cultural Heritage Preservation and Zoning of Perm, Russia

Scientific and Projecting Institute of Spatial Planning ENKO

St. Petersburg, Russia
By the ENKO stafff

Contact

Olga Krasovskaya
olgakrasovskaya@mail.ru

Software

ArcView 3.2, ArcView 8.2, Avenue, Linux RH, and Oracle9i database for Linux/IA64

Hardware

Intel PIV 2.0

Printer

HP Designjet 120 and HP Deskjet 1220

Data Source(s)

Department of Architecture and Planning

Conservation

Prioritizing Land Management in the Conservation of a Rare Species

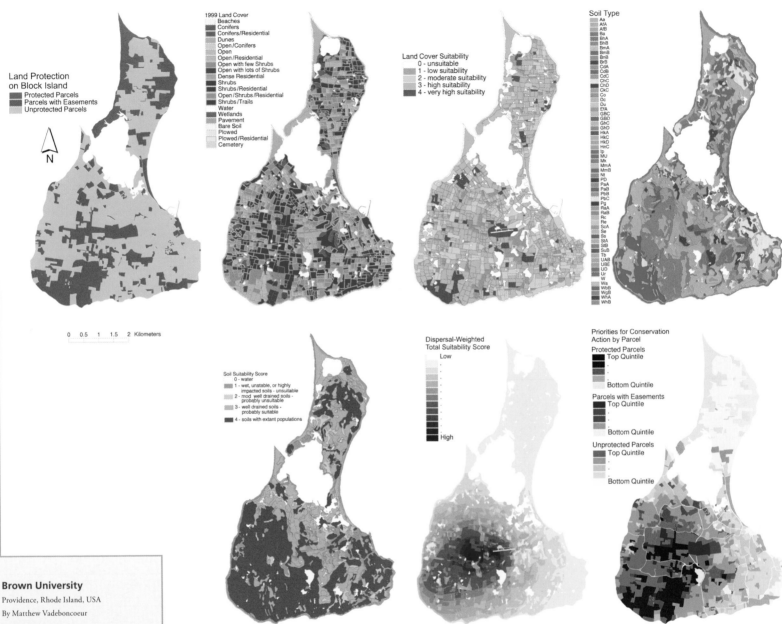

Brown University

Providence, Rhode Island, USA

By Matthew Vadeboncoeur

Contact

Matthew Vadeboncoeur

vadeboncoeur@alumni.brown.edu

Software

ArcView 3.2, Adobe Illustrator 10, and Windows

Hardware

Solaris

Printer

HP Designjet 755cm

Data Source(s)

Rhode Island Department of Environmental Management; Rhode Island GIS; and Town of New Shoreham, Rhode Island

Conservation

The work consists of an analysis of the current habitat of a rare wildflower and a model to predict where it might be reestablished to locate land parcels for potential acquisition and other conservation action.

New England blazing star (*Liatris scariosa* var. *novae-angliae*) is a rare grassland perennial species endemic to New England and New York, which occurs primarily along the New England coast. Twelve of Rhode Island's 13 element occurrences of New England blazing star occur on Block Island, Rhode Island. Some of these populations have fewer than 100 individuals and do not appear to be able to persist on their own.

If New England blazing star persists as a metapopulation on Block Island, then the conservation of both occupied and vacant habitat patches is important to the long-term persistence of the metapopulation. The author identified suitable habitat for New England blazing star using Rhode Island GIS data on soils and land use compiled from aerial photographs. He created a model of long distance dispersal based on an exponential decay function, weighted by the size of each population in 2000. Habitat suitability was combined with the dispersal kernel to create a map of relative colonization likelihood called a "total suitability score." The mean total suitability score was calculated for each parcel and used to prioritize individual parcels for conservation action. Appropriate conservation actions include landowner outreach and education, the active management of invasive and successional species, and the exclusion of deer.

Ecology and Empire Along the Ancient Silk Roads

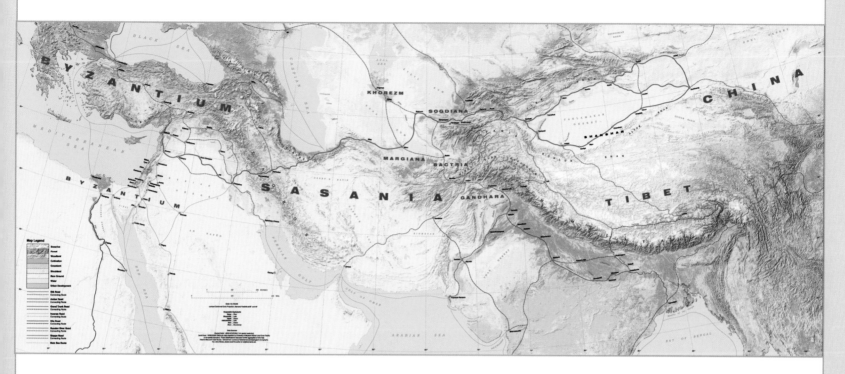

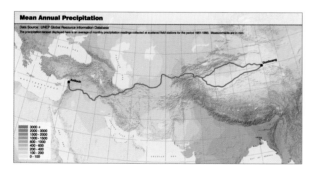

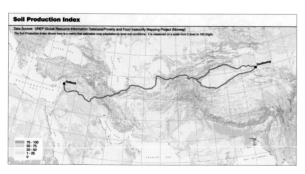

Before the era of mechanized travel, the silk roads were the human highways linking Asia and Europe. Across this vast network of trade routes, merchants plied their wares—silks, furs, gems, metals, and foods from every corner of the known world.

Transcontinental commerce reached its apex in the early seventh century A.D., when powerful and politically stable empires enabled trade to flourish. The demand for exotic goods that fueled this trade was underwritten by a single undeniable feature—the ecology of the Eurasian landmass.

The goal of this map is to provide a way to visualize the ecological context of the Silk Roads. The broad ecological zones of Asia and Europe—the steppes, deserts, montane forests, and alluvial floodplains—are approximated by modern land cover. Traversing these distinct ecological zones is a complex set of medieval trade routes that connected major cities and empires of the day. By examining the complex geography of these trade routes, the map seeks to enrich our understanding of the relationship between ecology, economy, and empire during one of the most important periods in human history.

EDAW
Seattle, Washington, USA
By Rob Harris

Contact
Rob Harris
rob@rob-harris.net

Software
ArcGIS 8.3 and Windows XP

Printer
HP Designjet 5500

Data Source(s)
National Oceanic Atmospheric Administration/National Aeronautics and Space Administration Pathfinder Program, United Nations Environment Programme, U.S. Geological Survey, and historical and archaeological monographs and reports

Conservation

Electrical Distribution Map, Elmendorf Air Force Base, Alaska

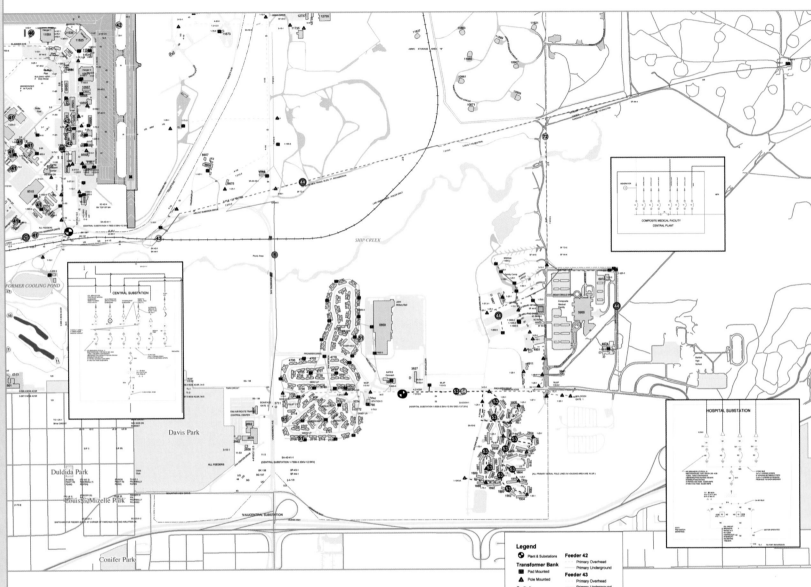

U.S. Air Force, Pacific Air Forces

Hickam Air Force Base, Alaska, USA

By Del Brown, Erik Jackson, and Ben McMillan

Contact

Ben McMillan

benjamin.mcmillan@elmendorf.af.mil

Software

ArcInfo 8.2 and Windows

Printer

HP Designjet 500

Data Source(s)

CAD drawings, geodatabases, and shapefiles

Defense and Intelligence

The U.S. Air Force GeoBase enabled the production of Primary Electrical Distribution maps with inserted schematic drawings for field use and planning. Dated cadastral data was used as reference for field surveys and verification using Trimble ProXRS GPS (submeter accuracy) and Trimble 5700 GPS (subcentimeter accuracy) units. Schematic drawings were added in separate data frames and placed in strategic layout areas for specific locations to highlight detailed information not captured at the produced scale of the overall map. Symbology was produced by referencing Tri-Services Spatial Data Standards and input from local electrical engineers.

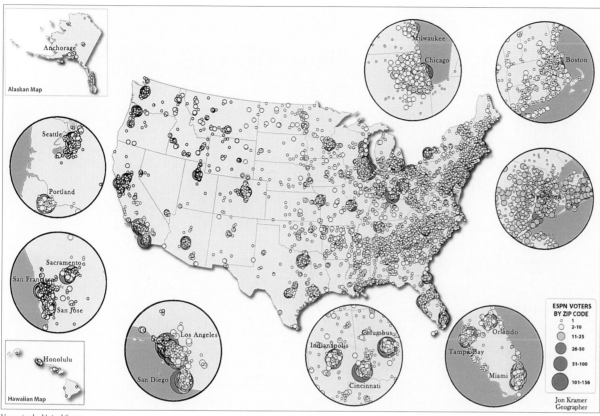

Voters in the United States

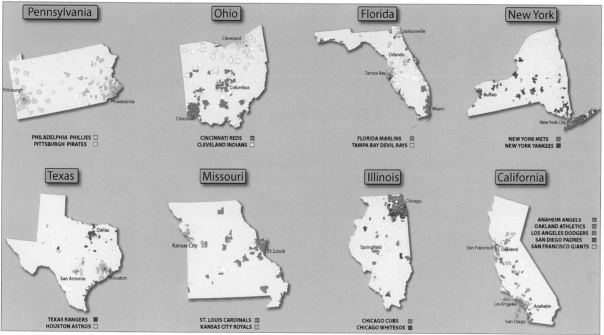

Major League Baseball

I n summer 2002, the Entertainment and Sports Programming Network (ESPN) conducted an online survey of more than 40,000 sports fans nationwide. ESPN called the survey ESPN's SportsNation and published some of the results in the August 5, 2002, edition of its magazine.

This work used ESPN's data from the survey to create maps showing geographical patterns in the voting results. In a survey of approximately 30 questions, the mapping possibilities are seemingly endless, leading to in-depth mapping analysis for specific sports-related topics. This poster presentation shows a sample of sports maps derived from the data.

University of Wisconsin–Eau Claire

Eau Claire, Wisconsin, USA

By Brady Foust and Sean Hartnett

Contact

Brady Foust

bfoust@uwec.edu

Software

ArcGIS 8.2

Hardware

Dell Pentium PC

Printer

HP Laserjet 4500

Data Source(s)

ESPN survey

Education

Quantification of Reservoir Potential of the Devonian Limestone East Waddell Ranch Survey, Texas

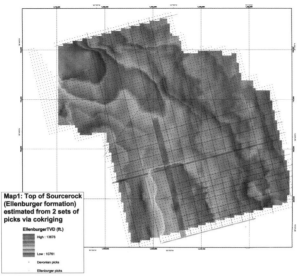

Map1: Top of Sourcerock (Ellenburger formation) estimated from 2 sets of picks via cokriging

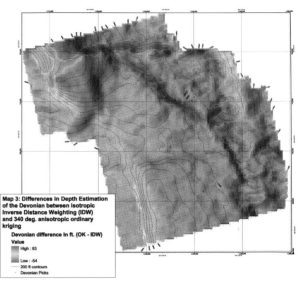

Map 3: Differences in Depth Estimation of the Devonian between isotropic Inverse Distance Weighting (IDW) and 340 deg. anisotropic ordinary kriging

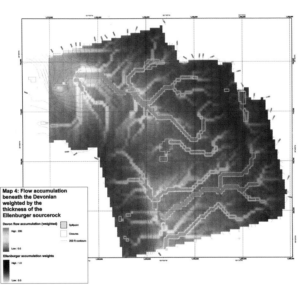

Map 4: Flow accumulation beneath the Devonian weighted by the thickness of the Ellenburger sourcerock

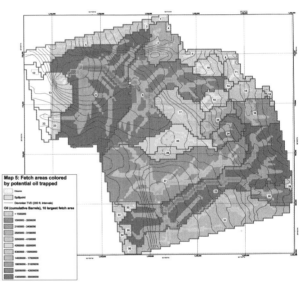

Map 5: Fetch areas colored by potential oil trapped

University of Houston–Geoscience Department

Houston, Texas, USA

By Chris Harding and Tracy Thorleifson

Contact

Chris Harding

charding@iastate.edu

Software

ArcGIS 8.3, ArcGIS Spatial Analyst, ArcGIS Geostatistical Analyst, and Windows XP

Printer

HP Designjet 1055cm

Data Source(s)

Point samples of two sediment layers manually picked from seismic data

Education

This display shows work done for a GIS course given at the University of Houston. Data from two sediment layers was used to quantify reservoir potential. Map 1 used geostatistics (anisotropic cokriging) to estimate the depth of the oil producing (lower) layer from point samples. Map 3 shows the depth of the oil trapping (upper) layer. The color shows differences between a simple estimation procedure (IWD) and kriging (OK). Map 4 shows a flow simulation of oil below the trapping layer toward the surface. Map 5 shows the fetch areas (basins) color coded by potentially trapped oil. The bar chart shows the most valuable fetch areas and their comparable extend.

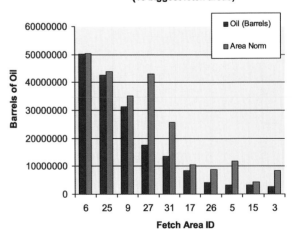

Maximum Oil per Fetch Areas
(10 biggest fetch areas)

Perchlorate—Drinking Water Impacts and Sources in California

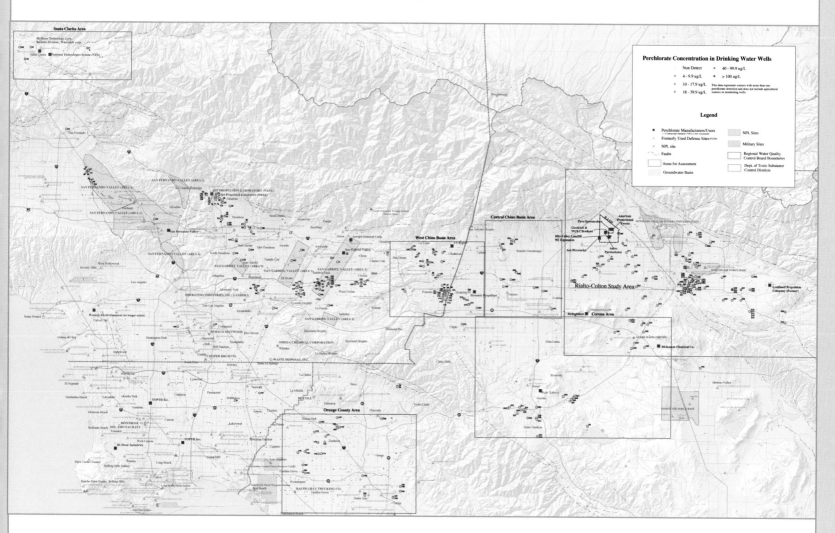

Perchlorate is a man-made salt that has recently been discovered as a threat to drinking water supplies. Perchlorate is a manufactured component of solid rocket fuel and is used for explosives, fireworks, flares, and other applications. In humans and animals, perchlorate can adversely affect the thyroid. Prior to April 1997, perchlorate could not be detected in low concentrations, and little was known about either its toxicity or the extent perchlorate had contaminated drinking water supplies. Then in 1997, the California Department of Health Services developed a new analytical method to detect very low levels of perchlorate in water. Since then, this chemical has been found in the water supplies of more than 15 million people in California, Nevada, and Arizona as well as in surface and groundwater throughout the United States.

This map brings together information from the state of California and several federal agencies to help prioritize and coordinate efforts to address this emerging contaminant. State agencies have been tracking public water supply wells where perchlorate has been detected. The U.S. Environmental Protection Agency (EPA) has information on many of the major sources such as aerospace manufacturing operations that are now Superfund cleanup sites. The Department of Defense has information on former military sites throughout California. Nine areas that have been identified by U.S. EPA for further investigation can be seen in the "perchlorate discovery study areas" inset. This map and future updates provide a useful and flexible tool for communication within and between government agencies.

U.S. Environmental Protection Agency Pacific Southwest Region/ Titan Corp.

San Francisco, California, USA

By Linda K. Chambers, Titan Corp.; Kevin Mayer and Matt Mitguard, U.S. EPA, Region 9

Contact

Kevin Mayer

mayer.kevin@epa.gov

Software

ArcInfo and Windows 2000

Printer

HP Designjet 5000

Data Source(s)

California Department of Health; U.S. Army Corps of Engineers; and U.S. EPA, Region 9

Environmental Management

National Oil Spill Contingency Plan for Cameroon—Vegetation Map

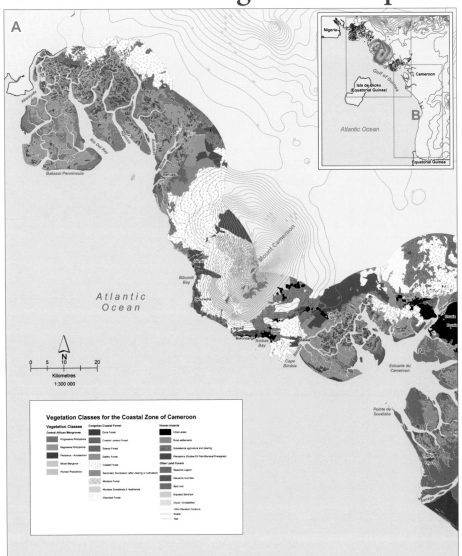

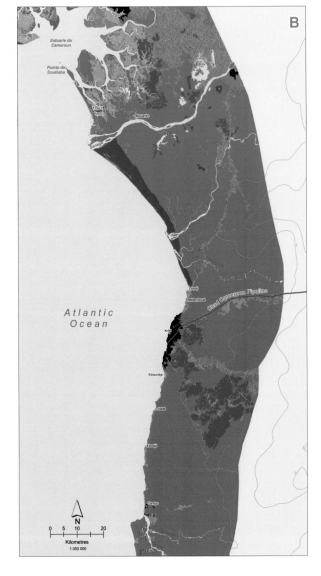

CSIR Environmentek

Stellenbosch, South Africa

By Simon Hughes and David Le Maitre

Contact

Simon Hughes

shughes@csir.co.za

Software

ArcMap 8.3 and ERDAS IMAGINE 8.6

Hardware

Dell Precision Workstation 650

Printer

HP Designjet 5500

Data Source(s)

Central Africa Regional Program for the Environment, National Geospatial-Intelligence Agency, and U.S. Geological Survey

---

Environmental Management

Following the recent construction of the 1,000-kilometer Chad–Cameroon oil pipeline, CSIR was commissioned by the World Bank to prepare a National Oil Spill Contingency Plan for Cameroon. Comprising the core of the plan is an appraisal of the various sources of oil spill risk to the environment and an analysis of environmental sensitivity to oil. Using Landsat 7 ETM+ satellite imagery, a map of the coastal vegetation of Cameroon was prepared as the basis for determining environmental sensitivity. Other ecological factors and geomorphological and socioeconomic criteria also informed the sensitivity analysis. The map depicts the main vegetation types that occupy Cameroon's littoral zone, highlighting those that could be affected either directly or indirectly by oil spills. It differentiates between four mangrove communities, freshwater swamp forest, two forms of dune forest, and a variety of transformed environments (e.g., urban and rural settlements). In the National Oil Spill Contingency Plan, these are described in terms of their relative sensitivity to contamination by oil spills.

The vegetation map is also a useful tool for general coastal zone management. It depicts a baseline state against which changes in distribution and extent of vegetation types and transformed areas can be measured over time. Trends can be quantified by using the baseline state shown on the vegetation map as a reference point.

Generalized Contours of the Sauk Sequence for Characterization of Saline Aquifers for CO$_2$ Sequestration

THICKNESS OF THE MOUNT SIMON SANDSTONE

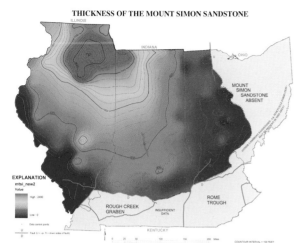

STRUCTURE ON THE MOUNT SIMON SANDSTONE

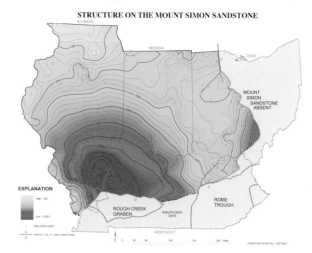

STRUCTURE ON THE KNOX UNCONFORMITY

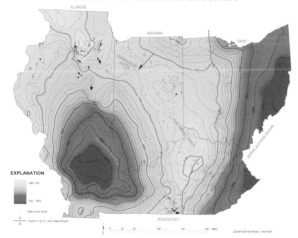

STRUCTURE ON THE PRECAMBRIAN UNCONFORMITY

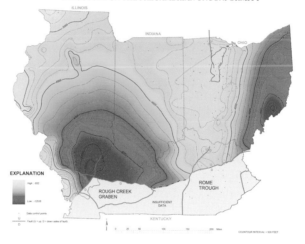

The Sauk sequence is the lowest major sequence of rocks (a largely uninterrupted depositional system, bounded below and above by major nonconformities [or periods of nondeposition]) within the Paleozoic strata of North America. The lowermost unit of this sequence is the Mount Simon Sandstone of Middle Cambrian Age. The Mount Simon Sandstone was deposited on a major nonconformity surface of eroded Precambrian Age strata. Within the southern portion of the mapped region, large Cambrian Age rift faults are responsible for dramatic accumulations of sedimentary rocks within the rift systems. Few wells have been drilled to appreciable depths within these rift sequences to allow characterization of their carbon dioxide (CO$_2$) sequestration potential. These areas are blank on the Precambrian and Mount Simon maps.

Mount Simon Sandstone is an excellent candidate reservoir for CO$_2$ sequestration. Favorable sequestration attributes of Mount Simon Sandstone include its relatively high porosity and permeability; its lateral continuity over a multistate area; its thickness; its occurrence at depth beneath thousands of feet of alternating sequences of carbonate, shale, and sandstone strata; and its stratigraphic position directly overlying relatively impermeable Precambrian Age strata. Mount Simon Sandstone underlies one of the most densely populated regions of the United States.

Previous interpretations of Mount Simon Sandstone have shown it to be present over most of Ohio and much of the Appalachian basin. However, current research at the Ohio Geological Survey indicates that Mount Simon Sandstone is not present in much of eastern and central Ohio. The basal stratum in these areas is mostly dolomite with interbeds of sandstone and sandy dolomite. This is the first publicly presented map showing the absence of Mount Simon Sandstone in central and eastern Ohio. More detailed maps on the occurrence and thickness of Mount Simon Sandstone will be made available via the MIDCARB Web site.

The Knox unconformity surface is the top of the Sauk sequence. Rocks of the Knox Group also are extensive, stretching across most of the interior of the North American continent. The Knox Group is largely composed of dolomite and dolomitic sandstone. In eastern and central Ohio, this interval is an important producer of oil and gas and may have potential for value-added sequestration of CO$_2$. The presence of paleokarst-related porosity systems within the Knox Group may prove to be an effective CO$_2$ reservoir in some places throughout the region. The horizon shown here mainly defines the top of the Sauk sequence—a 1,000- to 4,000-foot-thick interval of shale, siltstone, sandstone, dolomite, and limestone—that contains CO$_2$, which could leak from a Mount Simon CO$_2$ sequestration site.

Indiana University
Bloomington, Indiana, USA
By J.B. Hickman, C.P. Korose,
D.M. Powers, W. Solano, and
L.H. Wickstrom

Contact
Premkrishnan Radhakrishnan
pradhakr@indiana.edu

Software
ArcGIS 8.2, GeoGraphix, Microsoft
Access, Microsoft Excel, and Windows 2000

Printer
HP Designjet 5500ps

Data Source(s)
Illinois, Indiana, Kentucky, and Ohio
geological surveys

Environmental Management

Impact Mitigation Analysis of Helicopters on Mountain Goats in British Columbia

Goat Polygon Section

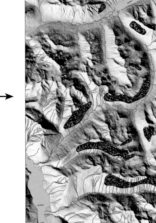

Random Point Generation

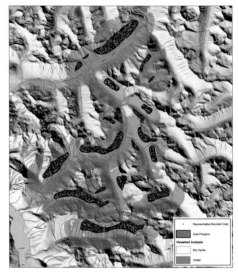

Results of Viewshed Analysis

The winter range of mountain goat *(Oreamnos americanus)* and helicopter-assisted backcountry recreation activities often occur in the same mountainous regions of British Columbia, Canada. Mountain goats have been identified as a provincially yellow-listed species of concern that are sensitive to helicopter disturbance. Cascade Environmental Resource Group Ltd. was contracted by Coast Range Heliskiing Ltd. to develop methodology for determining helicopter flight paths within its commercial recreation tenure, which eliminate visual and drastically minimize auditory disturbance to potential mountain goat populations. This poster illustrates the various stages of the analysis performed to reach that objective.

Critical winter range for the mountain goats within the commercial recreation tenure area was identified and populated with randomly generated points representing mountain goats throughout the range at a density of one per hectare. Attributes were added to each point to include a stationary rotation point, a radial distance for analysis of two kilometers (in accordance with government recommendations), a vertical offset to account for the standing height of a mountain goat, and a second vertical offset to account for flying height of helicopters above ridge lines. A viewshed analysis was performed incorporating the above parameters and using a base triangulated irregular network generated from 20-meter contours. Helicopter flight lines, and pickup/drop-off locations were subsequently adjusted to avoid regions visible to potential mountain goat populations. Where logistical flight paths are required within the determined goat viewshed, helicopters are required to maintain an elevation of 2,000 meters above the surface.

The study is based on the premise that terrestrial barriers, such as mountain ridges, influence what an individual animal is able to see and hear. This technique could be applied to mitigation strategies for a variety of sensitive species in land management situations in which terrestrial barriers are a factor.

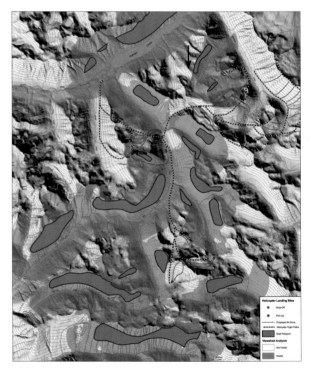

Mitigated Helicopter Flight Paths

Cascade Environmental Resource Group Ltd.

Whistler, British Columbia, Canada
By Kristen Harrison, Chris McDougall, and Dave Williamson

Contact
Dave Williamson
dwilliamson@cascade-environmental.ca

Software
ArcGIS 8.3, ArcGIS 3D Analyst, and Windows XP

Hardware
Pentium 4

Printer
HP Designjet 755cm

Data Source(s)
British Columbia, British Columbia Geological Survey, and Ministry of Sustainable Resource Management

Environmental Management

Naval Aircraft in Hampton Roads, Virginia

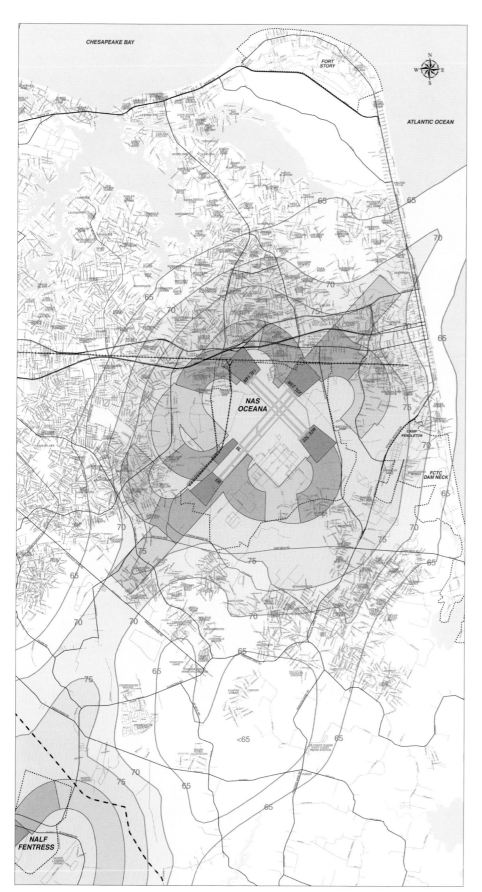

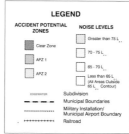

I n the early 1970s, the U.S. Department of Defense (DoD) established the Air Installations Compatible Use Zones (AICUZ) Program to balance the need for aircraft operations and community concerns. The goal of the AICUZ Program is to protect the health, safety, and welfare of those living near a military airport while preserving its defense flying mission.

Under the AICUZ Program, DoD provides noise zones and Accident Potential Zones (APZs) as planning tools for location planning agencies and recommends compatible development near military air facilities. Ecology and Environment, Inc., designed this map for public distribution to clearly show the noise zone levels and APZs affecting neighborhoods surrounding three naval air stations in Virginia.

LEGEND

ACCIDENT POTENTIAL ZONES

- Clear Zone
- APZ 1
- APZ 2

NOISE LEVELS

- Greater than 75 L$_{dn}$
- 70 - 75 L$_{dn}$
- 65 - 70 L$_{dn}$
- Less than 65 L$_{dn}$ (All Areas Outside 65 L$_{dn}$ Contour)

Subdivision
Municipal Boundaries
Military Installation/ Municipal Airport Boundary
Railroad

Ecology and Environment, Inc.

Lancaster, New York, USA

By Joe Ghosen

Contact

Lisa Matthies

lmatthies@ene.com

Software

ArcInfo and ArcPlot™

Printer

HP Designjet 1055cm+

Data Source(s)

Ecology and Environment, Inc.

Environmental Management

Yemen Country Profile

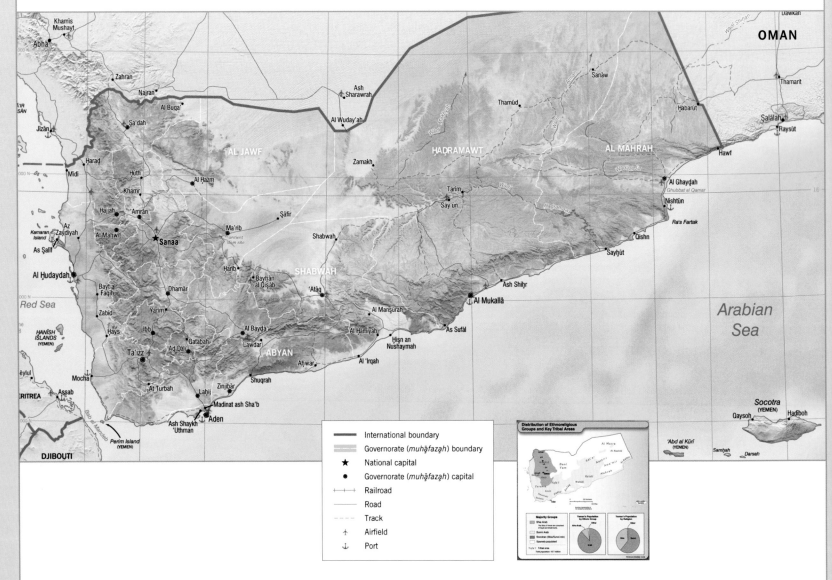

Legend:
- International boundary
- Governorate (*muhāfazah*) boundary
- ★ National capital
- ● Governorate (*muhāfazah*) capital
- ┼┼┼ Railroad
- Road
- Track
- ⟰ Airfield
- ⚓ Port

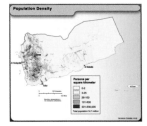

Distribution of Ethnoreligious Groups and Key Tribal Areas

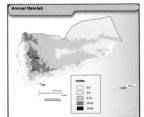

Population Density

Annual Rainfall

Economic Activity

Land Use

The Yemen Country Profile is part of a series of maps created to satisfy the need for quick turnaround reference products that address key geographic issues with high intelligence value for selected countries.

The main map consists of a large-format reference map, an orthographic distance comparison map, and an area comparison map. The remaining elements include a three-dimensional model highlighting the physical geography of the country, an ethnoreligious and key tribal areas map, a population density map, an annual rainfall map, an economic activity map, a land use map, a timeline describing the major events that have transpired since 1959, and key facts taken from the CIA's *World Factbook*.

U.S. Government

Washington, D.C., USA

Central Intelligence Agency

Software

ArcView 3.2a, Adobe Illustrator, Adobe Photoshop, Corel Bryce, ERDAS IMAGINE, MAPublisher, and Windows NT

Printer

HP Designjet

Data Source(s)

Central Intelligence Agency, National Geospatial-Intelligence Agency, and remote imagery

Government—Federal

Sex Offender Proximity Analysis

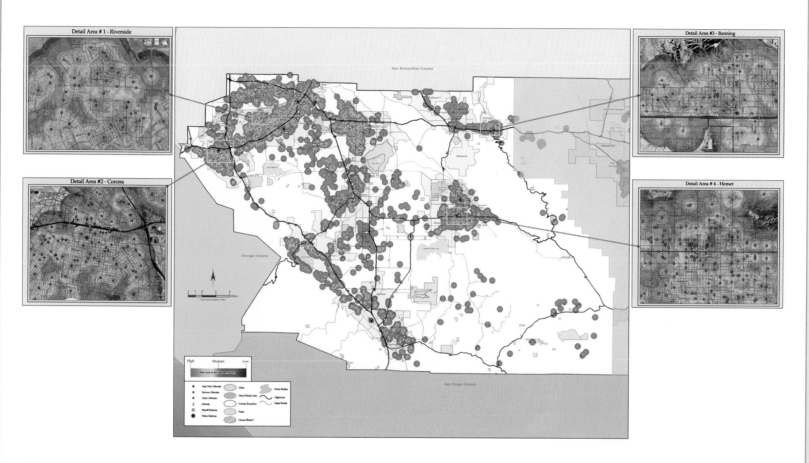

The purpose of this analysis was to determine the high-risk areas within Riverside County where registered sex offenders could threaten children. To obtain this information, four different data sets were combined to locate these areas. The data sets included addresses of registered sex offenders, census block information, and locations of schools and parks.

Data on the addresses of known sex offenders was buffered at one-half mile intervals up to five miles. Those intervals were reclassified into zones 1 through 10. Ten represents the closest zone to the location of the sex offender and one the farthest away.

The census block data set was reselected for the median age of 18 and under. Of the census blocks selected, the population under the age of 18 was 50 percent. This data was also buffered at one-half mile intervals up to five miles. The zones were reclassified in the same manner as the sex offender addresses. The schools and parks data sets were similarly buffered and reclassified.

When the data was reclassified, the raster calculator in ArcGIS Spatial Analyst combined the information and weighted it as follows: sex offenders, 60 percent; schools, 30 percent; parks, 5 percent; and census block, 5 percent. The result of this calculation showed the most probable spot in Riverside County where a possible sex offense might occur against a minor.

Riverside County Transportation and Land Management Agency GIS
Riverside, California, USA
By Robert Conrad

Contact
Robert Conrad
rwconrad@co.riverside.ca.us

Software
ArcGIS Spatial Analyst, ArcInfo, Visio, and Windows 2000

Printer
HP Designjet 1055cm+

Data Source(s)
ESRI, State of California, Riverside County Sheriff, and U.S. Department of Justice

Government—Law Enforcement and Criminal Justice

2002 Seismic Hazard Maps for the Conterminous United States

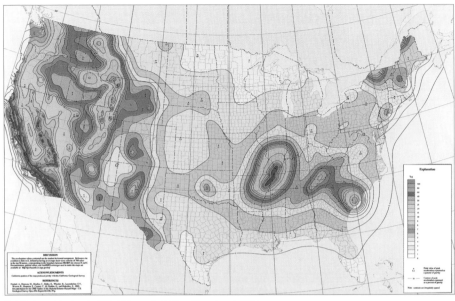

Peak Horizontal Acceleration With 10 Percent Probability of Exceedance in 50 Years

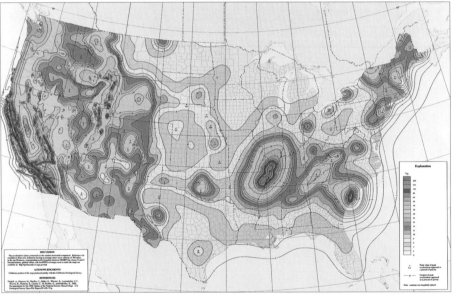

Peak Horizontal Acceleration With 2 Percent Probability of Exceedance in 50 Years

U.S. Geological Survey

Lakewood, Colorado, USA

By Ken Rukstales

Contact

Ken Rukstales

rukstales@usgs.gov

Software

ArcGIS 8.2 Workstation, ArcInfo, and Windows 2000

Printer

HP Designjet 5000

Data Source(s)

U.S. Geological Survey

Government—Public Safety

National maps of earthquake shaking hazards provide information essential to creating and updating the seismic design provisions of building codes used in the United States. Scientists frequently revise these maps to reflect new knowledge. The U.S. Geological Survey National Seismic Hazard Mapping program has recently updated the seismic hazard maps for the United States. Buildings, bridges, highways, and utilities built to meet modern seismic design provisions are better able to withstand earthquakes not only saving lives but also enabling critical activities to continue with less disruption. Online, interactive versions of the maps, as well as seismicity and fault data used to produce the maps, are available at www.eqmaps.cr.usgs.gov. Documentation, gridded values, and ArcInfo coverages used to make the maps are at http://geohazards.cr.usgs.gov/eq/.

Earthquake Shaking Intensities for California

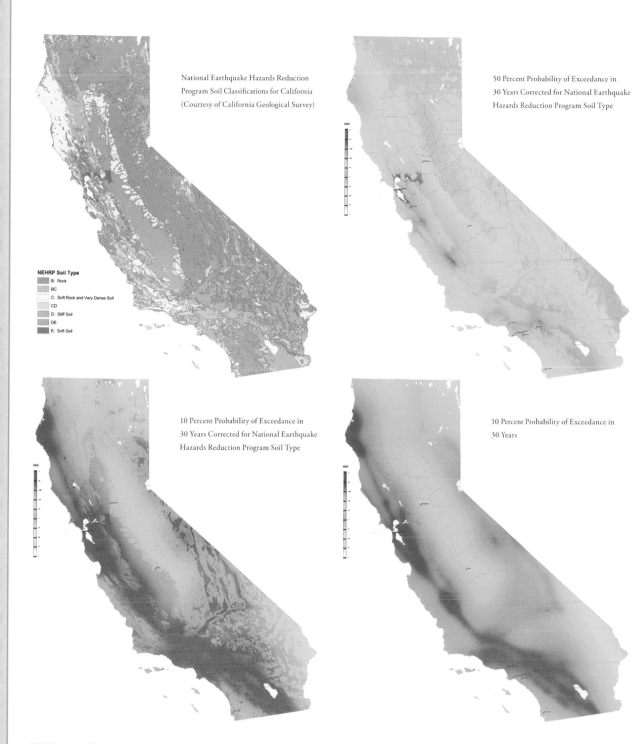

National Earthquake Hazards Reduction Program Soil Classifications for California (Courtesy of California Geological Survey)

NEHRP Soil Type
- B: Rock
- BC
- C: Soft Rock and Very Dense Soil
- CD
- D: Stiff Soil
- DE
- E: Soft Soil

50 Percent Probability of Exceedance in 30 Years Corrected for National Earthquake Hazards Reduction Program Soil Type

10 Percent Probability of Exceedance in 30 Years Corrected for National Earthquake Hazards Reduction Program Soil Type

10 Percent Probability of Exceedance in 30 Years

Expected levels of earthquake shaking for a region can be calculated by combining knowledge of the likelihood and magnitude of future earthquakes and rupture of specific fault segments with information on how seismic waves propagate through the region and how they affect local soil conditions.

The U.S. Geological Survey and the California Geological Survey have jointly produced seismic hazard maps for California using this knowledge. These maps show the likelihood of exceeding a given level of shaking during a specific time interval. The expected levels of ground motion can be converted to earthquake intensity levels that are directly related to expected damage.

The authors calculated two intensity hazard maps for a 30-year time window. One map shows the probability that an area has a 1 in 10 (10 percent) chance of intensity being exceeded in 30 years. The other map shows areas having even odds (50 percent) of intensity being exceeded in 30 years. Local soil conditions play a significant role in increasing expected intensities in areas of soft sediment and fill.

U.S. Geological Survey
Lakewood, Colorado, USA
By Ken Rukstales and Mark Petersen

Contact
Ken Rukstales
rukstales@usgs.gov

Software
ArcGIS 8.2 and Windows 2000

Printer
HP Designjet 5000

Data Source(s)
U.S. Geological Survey

Government—Public Safety

City of El Cajon Fire Department Midday Response Times

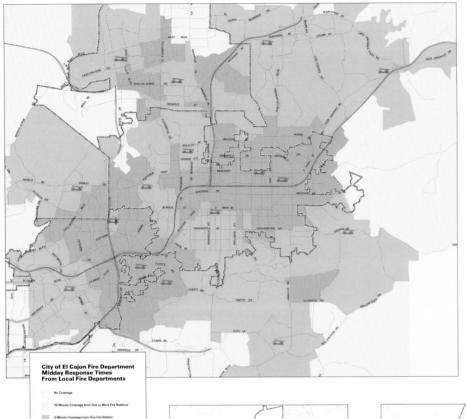

City of El Cajon Fire Department
Midday Response Times
From Local Fire Departments

- No Coverage
- 10 Minute Coverage from One or More Fire Stations
- 4 Minute Coverage from One Fire Station
- 4 Minute Coverage from Two or More Fire Stations
- Fire Stations

The city of El Cajon has 16 fire stations within its sphere (6 shown). Travel times were obtained from the San Diego Association of Governments (SANDAG) travel demand model to be used for travel time analysis. The off-peak traffic assignment was used to mimic midday travel conditions. The department requested two-minute time intervals up to 10 minutes. Sixteen maps were prepared to show each fire station site independently. A larger map shows the overlap analysis of all 16 fire stations combined. The fire department will use these maps for updating response areas for emergency services and the accreditation process.

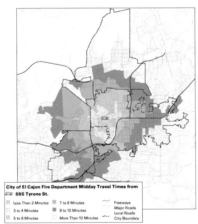

City of El Cajon Fire Department Midday Travel Times from
695 Tyrone Blvd.

- Less Than 2 Minutes
- 3 to 4 Minutes
- 5 to 6 Minutes
- 7 to 8 Minutes
- 9 to 10 Minutes
- More Than 10 Minutes
- Freeways
- Major Roads
- Local Roads
- City Boundary

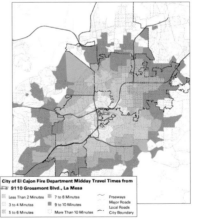

City of El Cajon Fire Department Midday Travel Times from
9110 Grossmont Blvd., La Mesa

- Less Than 2 Minutes
- 3 to 4 Minutes
- 5 to 6 Minutes
- 7 to 8 Minutes
- 9 to 10 Minutes
- More Than 10 Minutes
- Freeways
- Major Roads
- Local Roads
- City Boundary

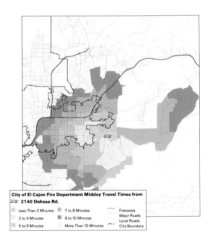

City of El Cajon Fire Department Midday Travel Times from
2140 Dehesa Rd.

- Less Than 2 Minutes
- 3 to 4 Minutes
- 5 to 6 Minutes
- 7 to 8 Minutes
- 9 to 10 Minutes
- More Than 10 Minutes
- Freeways
- Major Roads
- Local Roads
- City Boundary

San Diego Association of Governments

San Diego, California, USA

By Mike Calandra

Contact

Mike Calandra

mca@sandag.org

Software

ArcInfo 8.2 Workstation and UNIX

Printer

HP Designjet 5000ps

Data Source(s)

SANDAG and SanGIS

Government—Public Safety

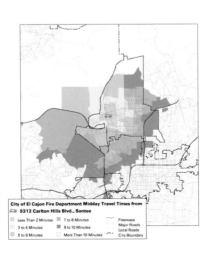

City of El Cajon Fire Department Midday Travel Times from
9312 Carlton Hills Blvd., Santee

- Less Than 2 Minutes
- 3 to 4 Minutes
- 5 to 6 Minutes
- 7 to 8 Minutes
- 9 to 10 Minutes
- More Than 10 Minutes
- Freeways
- Major Roads
- Local Roads
- City Boundary

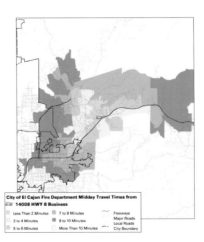

City of El Cajon Fire Department Midday Travel Times from
14008 HWY 8 Business

- Less Than 2 Minutes
- 3 to 4 Minutes
- 5 to 6 Minutes
- 7 to 8 Minutes
- 9 to 10 Minutes
- More Than 10 Minutes
- Freeways
- Major Roads
- Local Roads
- City Boundary

City of El Cajon Fire Department Midday Travel Times from
9726 Riverview St.

- Less Than 2 Minutes
- 3 to 4 Minutes
- 5 to 6 Minutes
- 7 to 8 Minutes
- 9 to 10 Minutes
- More Than 10 Minutes
- Freeways
- Major Roads
- Local Roads
- City Boundary

2003 North East/Gippsland Fires

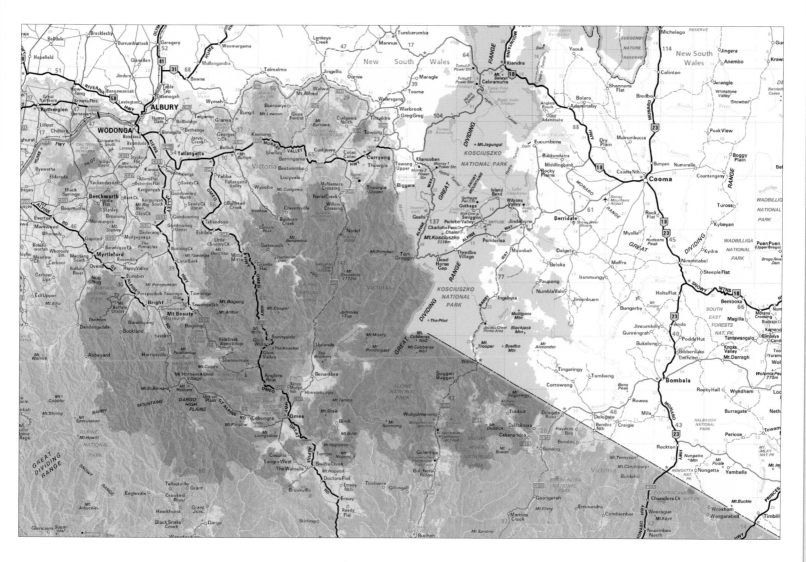

After an extended period of drought, on January 7, 2003, dry lightning strikes ignited a series of bushfires that gradually consumed more than 1,900,000 hectares including towns in the alpine areas of Victoria and New South Wales during an eight-week period.

GIS technology played a significant role in the fire service's ability to understand the bushfires. It enhanced decision making both at the local level on the fire lines and at the strategic level where statewide resource issues could be managed effectively. Knowing where fire boundaries and active fire edges are located over such a large area was a huge bonus to the planning of the fire management operations. The public and partners in the fire industry, such as forestry and water companies, were kept informed of the fire situation via a suite of mapping products on the Internet.

Country Fire Authority of Victoria
Victorian Department of Sustainability and Environment
Melbourne, Victoria, Australia
By Mark Garvey

Contact
Mark Garvey
m.garvey@cfa.vic.gov.au

Software
ArcMap

Printer
Offset printer

Data Source(s)
Country Fire Authority of Victoria, Victorian Department of Sustainability and Environment; Melway Publishing Pty. Ltd.; New South Wales Rural Fire Service; and Thematic Mapper satellite imagery

Government—Public Safety

City of Berkeley HAZUS Implementation

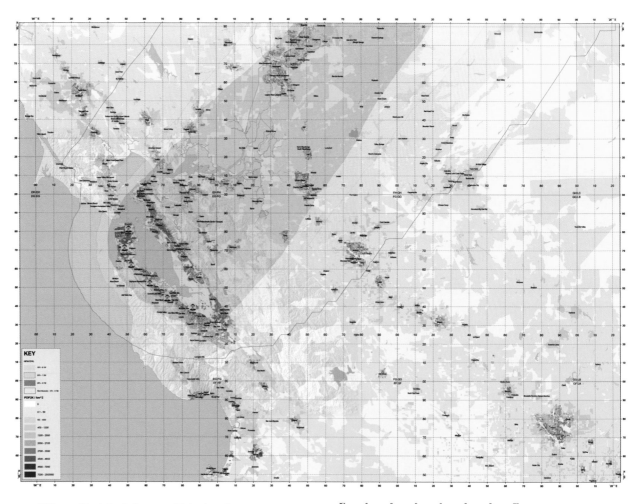

Ballistic Strike Fallout Trajectory

City of Berkeley

Berkeley, California, USA

By Brian B. Quinn

Contact

Brian Quinn

bbq@ci.berkeley.ca.us

Software

ArcGIS 8.3, ArcView, ERDAS
IMAGINE 8.6, and Windows 2000

Hardware

Xeon

Printer

HP Designjet 1055cm

Data Source(s)

City of Berkeley, California Geological
Survey, California Spatial Information
Library, Federal Emergency Management
Agency, and U.S. Geological Survey

Government—Public Safety

City of Berkeley Information Technology Department, Geographic Information Systems Division, has used loss estimation and consequences assessment tools to support emergency response training exercises and disaster mitigation planning. These maps highlight some of the natural and technological hazards that were considered during 2003 and are shown on a base map emphasizing Census 2000 block residential population density.

Ballistic Strike Fallout Trajectory

This poster shows estimated mortality probability from six large thermonuclear warheads targeting the San Francisco Peninsula that was completed using an ArcView software-based consequence model linked to stratospheric wind data, which was gridded from observations near the scenario time. Lethal fallout reaches Sacramento, and an estimated seven to nine million lives are claimed.

Ground Motion Estimates—North Hayward Fault

City of Berkeley HAZUS Implementation

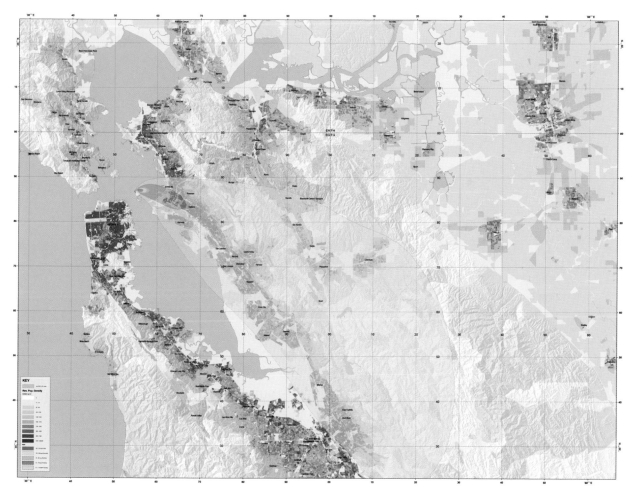

Radiological Weapon on Treasure Island

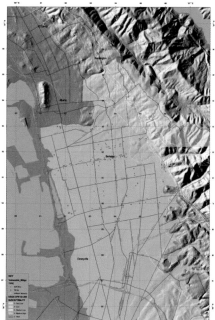

Ground Motion Estimates—North Hayward Fault

Radiological Weapon on Treasure Island

Estimated radiological dispersal from a device on Treasure Island in San Francisco Bay is shown. This analysis assessed consequences with an ArcView software-based model linked to surface winds, which was gridded from observations near the scenario time. The largest plume estimates a 10-millirem dose, which is small. For comparison, pregnant radiation workers are limited to fetal exposures of less than 50 millirem per month.

Ground Motion Estimates—North Hayward Fault

The Federal Emergency Management Agency's Natural Hazard Loss Estimation tool (HAZUS) was used to estimate ground motion for a Hayward Fault earthquake scenario as shown on the facing page. The analysis used U.S. Geological Survey (USGS) sources of seismic landslide susceptibility, liquefaction susceptibility as shown at left, and California Geological Survey soils seismic amplification data. These maps include overlays of USGS geologic line features and city of Berkeley vulnerable building inventory—unreinforced masonry, soft story, and tilt up structures. Ground motion was estimated on a 10-meter grid when USGS soils seismic amplifications were released.

City of Berkeley
Berkeley, California, USA
By Brian B. Quinn

Contact
Brian Quinn
bbq@ci.berkeley.ca.us

Software
ArcGIS 8.3, ArcView, ERDAS
IMAGINE 8.6, and Windows 2000

Hardware
Xeon

Printer
HP Designjet 1055cm

Data Source(s)
City of Berkeley, California Geological Survey, California Spatial Information Library, Federal Emergency Management Agency, and USGS

Government—Public Safety

Estimated Snowfall and Snow Load for Colorado— Storm of March 16–20, 2003

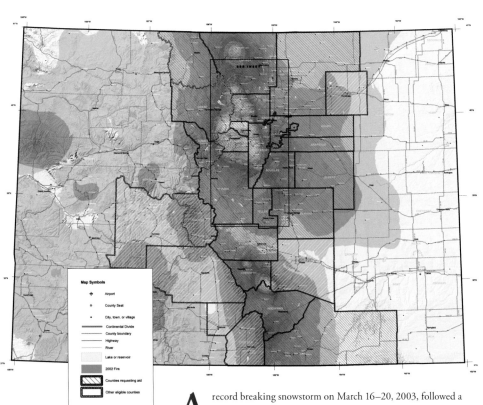

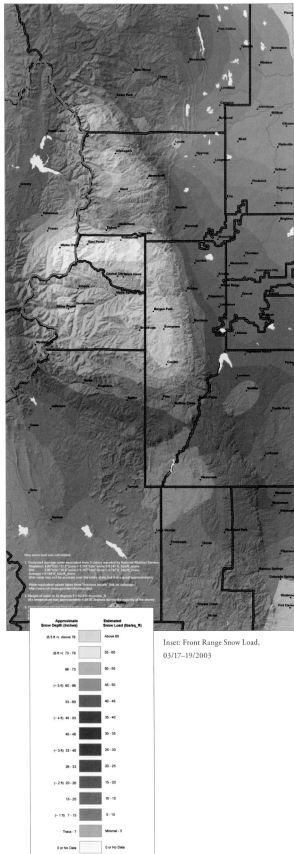

U.S. Geological Survey

Denver, Colorado, USA

By David Bortnem, Mark Feller,
John Kosovich, and Dick Vogel

Contact

John Kosovich
jjkosovich@usgs.gov

Software

ArcGIS Spatial Analyst, ArcInfo 8.2
Workstation, Microsoft Access, and
Windows 2000

Hardware

Dell

Printer

HP Designjet 5000

Data Source(s)

Colorado Department of Local Affairs,
Colorado Department of Transportation,
Colorado State University Community
Collaborative Rain and Hail Study, Federal
Emergency Management Agency, National
Oceanic and Atmospheric Administration/
National Weather Service, and
U.S. Geological Survey

Government—Public Safety

A record breaking snowstorm on March 16–20, 2003, followed a record breaking 2002 wildfire season for the state of Colorado. The Front Range and other eastern slope areas received several feet of snow, with many locations reporting five to six feet. With the help of this storm, which broke a streak of 19 consecutive months of below-normal precipitation, March 2003 became the snowiest March and the third snowiest month ever in Denver's recorded history.

Estimated snowfall totals on this map are based on the accumulations reported by the National Oceanic Atmospheric Administration/ National Weather Service and other weather spotters for this single storm. Snow depths at specific locations were obtained from the respective Web sites, interpolated into a depth surface, and combined with updated Colorado state base GIS data layers. Highlighted are counties that applied or were eligible for disaster aid. Also shown are locations of wildfires from the 2002 fire season. The map indicates that the heaviest snowfall occurred to the north of land burned by the 137,000-acre Hayman fire; but concerns of flooding and reservoir contamination due to spring melt-off throughout the stricken area existed well into the summer.

No accuracy is implied for the indicated snowfall depths, as the statewide depth surface was interpolated from relatively few (175) reporting sites falling mostly within the Front Range and eastern slope areas and because snow depths varied locally around each individual site. Snow load amounts also have no implied accuracy, because in addition to using estimated snow depths, load also uses an average snow water equivalent calculated from only two locations in Denver.

Inset: Front Range Snow Load, 03/17–19/2003

46

Official Zoning Map, City of Roswell, Georgia

Historic District Inset

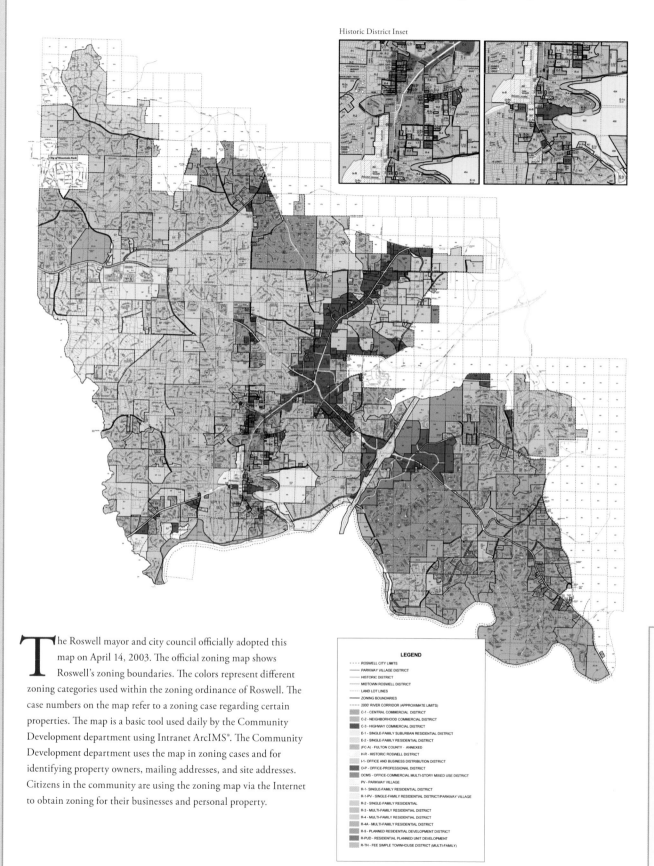

The Roswell mayor and city council officially adopted this map on April 14, 2003. The official zoning map shows Roswell's zoning boundaries. The colors represent different zoning categories used within the zoning ordinance of Roswell. The case numbers on the map refer to a zoning case regarding certain properties. The map is a basic tool used daily by the Community Development department using Intranet ArcIMS®. The Community Development department uses the map in zoning cases and for identifying property owners, mailing addresses, and site addresses. Citizens in the community are using the zoning map via the Internet to obtain zoning for their businesses and personal property.

LEGEND

- ROSWELL CITY LIMITS
- PARKWAY VILLAGE DISTRICT
- HISTORIC DISTRICT
- MIDTOWN ROSWELL DISTRICT
- LAND LOT LINES
- ZONING BOUNDARIES
- 2000 RIVER CORRIDOR (APPROXIMATE LIMITS)
- C-1 - CENTRAL COMMERCIAL DISTRICT
- C-2 - NEIGHBORHOOD COMMERCIAL DISTRICT
- C-3 - HIGHWAY COMMERCIAL DISTRICT
- E-1 - SINGLE-FAMILY SUBURBAN RESIDENTIAL DISTRICT
- E-2 - SINGLE-FAMILY RESIDENTIAL DISTRICT
- (FCA) - FULTON COUNTY - ANNEXED
- H-R - HISTORIC ROSWELL DISTRICT
- I-1 - OFFICE AND BUSINESS DISTRIBUTION DISTRICT
- O-P - OFFICE-PROFESSIONAL DISTRICT
- OCMS - OFFICE-COMMERCIAL MULTI-STORY MIXED USE DISTRICT
- PV - PARKWAY VILLAGE
- R-1 - SINGLE-FAMILY RESIDENTIAL DISTRICT
- R-1-PV - SINGLE-FAMILY RESIDENTIAL DISTRICT/PARKWAY VILLAGE
- R-2 - SINGLE-FAMILY RESIDENTIAL
- R-3 - MULTI-FAMILY RESIDENTIAL DISTRICT
- R-4 - MULTI-FAMILY RESIDENTIAL DISTRICT
- R-4A - MULTI-FAMILY RESIDENTIAL DISTRICT
- R-5 - PLANNED RESIDENTIAL DEVELOPMENT DISTRICT
- R-PUD - RESIDENTIAL PLANNED UNIT DEVELOPMENT
- R-TH - FEE SIMPLE TOWNHOUSE DISTRICT (MULTI-FAMILY)

City of Roswell
Roswell, Georgia, USA
By Sean D. Hamby and Derrick M. Smith

Contact
Derrick Smith
dsmith01@ci.roswell.ga.us

Software
ArcGIS 8.1, ArcGIS 8.2, ArcGIS 8.3, and ArcIMS

Hardware
Dell Workstation 330

Printer
HP Designjet 5000

Data Source(s)
City of Roswell

Government—State and Local

Yakima Traffic Maps

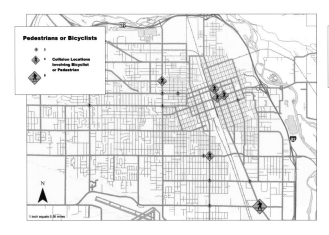

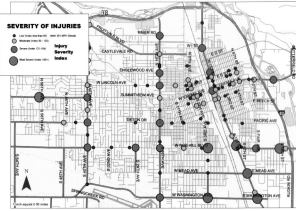

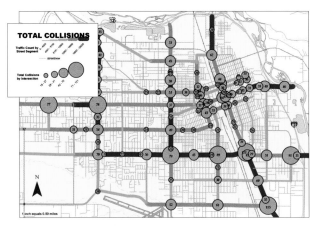

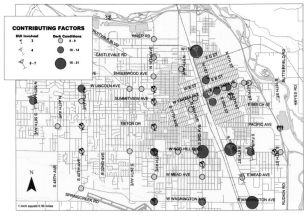

Traffic Safety Maps

Traffic safety is a measure of both a community's quality of life and infrastructure. Maps indicating the location and characteristics of vehicle collisions provide a primary source of information for traffic safety studies and identification of corrective measures to improve streets and intersections. Comparative information about different locations within the city is imperative in prioritizing construction and enhancement projects. Enforcement efforts from the police department are assisted by the quantitative and qualitative information shown on these maps.

These maps illustrate the traffic collisions within Yakima, Washington, during a five-year period—1998 through 2002. Data for these maps was collected from the Yakima Police Traffic Accident Field Reports. Yakima uses a software reporting tool, Intersection Magic, by Pd' Programming Products.

City of Yakima

Yakima, Washington, USA
By Jill Ballard, Joan Davenport, and Tom Sellsted

Contact

Tom Sellsted
tsellste@ci.yakima.wa.us

Software

ArcInfo 8.3 and Windows 2000

Printer

Xerox DocuColor 12 and
HP Designjet 755cm

Data Source(s)

Local data sources

Government—State and Local

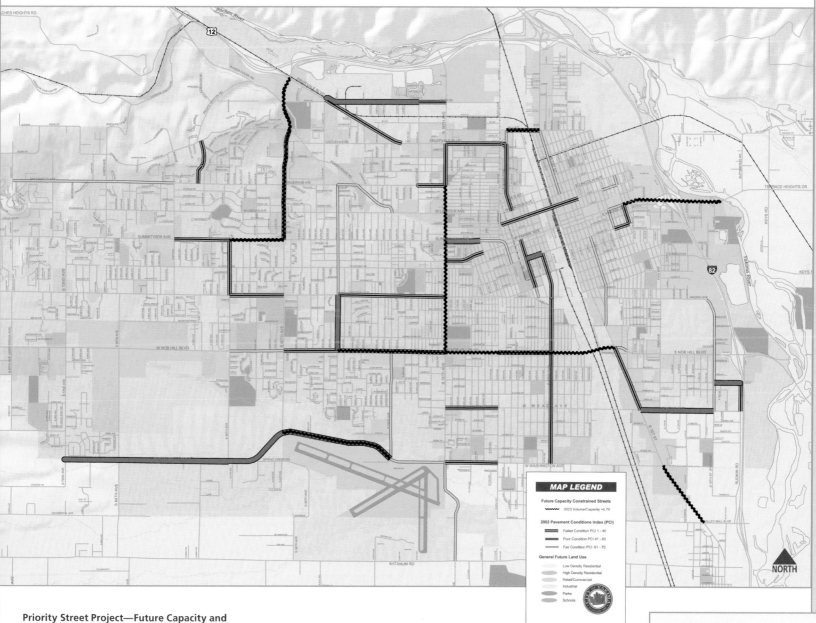

MAP LEGEND

Future Capacity Constrained Streets
- 2023 Volume/Capacity >0.70

2002 Pavement Conditions Index (PCI)
- Failed Condition PCI 1 - 40
- Poor Condition PCI 41 - 60
- Fair Condition PCI 61 - 70

General Future Land Use
- Low Density Residential
- High Density Residential
- Retail/Commercial
- Industrial
- Parks
- Schools

NORTH

Priority Street Project—Future Capacity and Severe Pavement Conditions

This map identifies arterial streets that require future improvement based on two separate conditions—future capacity constraints or severe pavement deficiencies. Projects identified on this map, as well as others submitted by the public participation process, require financial analysis and a needs assessment.

Yakima, Washington, is currently updating its Transportation Plan as required by the Washington State Growth Management Act. The ability of public streets to support future traffic is a basic guideline of the growth management process called "concurrency," meaning that street capacity must be concurrent with development. To plan for future capacity needs, each city establishes and adopts the local standards for concurrency.

The street segments identified on this map are critical improvement projects for Yakima. The public participation process will help to prioritize these projects and could identify other areas of concern. Each project must also be analyzed for economic and possible environmental impacts. Future capacity-constrained streets may require additional travel lanes or other measures to reduce congestion. In some cases, improving alternate routes could relieve the congestion on certain street segments. Significant public investment is required when a street's structural surface has failed or is near the failure point. Sometimes the street segment can be rehabilitated. Others need a complete rebuild of the road base and surface.

City of Yakima

Yakima, Washington, USA
By Jill Ballard, Joan Davenport, and Tom Sellsted

Contact

Tom Sellsted
tsellste@ci.yakima.wa.us

Software

ArcInfo 8.3 and Windows 2000

Printer

Xerox DocuColor 12 and HP Designjet 755cm

Data Source(s)

Local data sources

Government—State and Local

Portland Neighborhood Commercial Market Area Comparisons

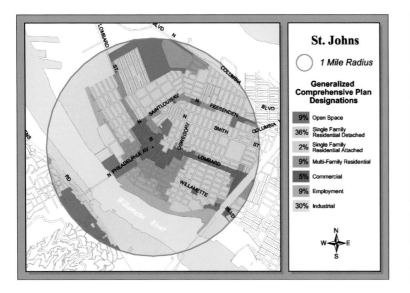

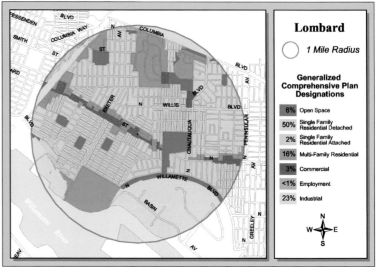

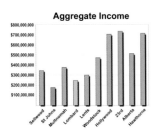

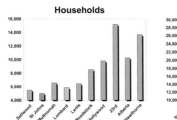

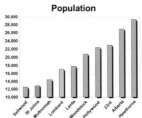

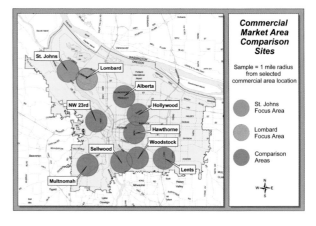

Bureau of Planning, City of Portland
Portland, Oregon, USA
By Carmen G. Piekarski

Contact
Carmen G. Piekarski
cpiekarski@ci.portland.or.us

Software
ArcInfo 8.2, ArcMap 8.2, Microsoft Excel, and Windows 2000

Hardware
Windows NT Workstation

Printer
HP Designjet 1055cm

Data Source(s)
City of Portland and U.S. Census Bureau

Government—State and Local

The St. Johns/Lombard Plan is a comprehensive area plan for the St. Johns town center and North Lombard main street areas in Portland, Oregon. One of the key goals of the plan is revitalization of the St. Johns and Lombard Street commercial areas. Commercial areas in St. Johns and along Lombard Street appear to be underperforming. There are vacant storefronts, and the range of goods and services being offered is limited and does not meet neighborhood needs and desires.

Market area characteristics, including population, income, and household size and composition, affect the performance of commercial districts. To better understand this relationship, the Portland Bureau of Planning conducted a comparison study of several Portland neighborhood commercial districts. For the comparison, a one-mile radius circle was drawn around each commercial center to approximate the area's local or primary market area. Demographic information and land use and transportation data from each area within the circle were collected and analyzed.

The study revealed differences between areas in factors that may affect commercial district success. The comparison showed that the St. Johns and Lombard Street areas have less population than many of the other places. The two areas also have lower median household incomes and larger household sizes than do several others. As a result, the St. Johns and Lombard Street local market areas may have less buying power and potentially different spending habits than some other areas. These factors could contribute to the type and amount of services offered in the community. In addition to local market characteristics, other factors, such as market image and type and mix of uses, contribute to commercial district success.

To foster commercial district revitalization, the plan's implementation strategy includes land use changes to enhance the market area. The plan provides more opportunities for housing in and near the retail core areas and in nearby residential areas. It also offers an opportunity to further expand the market area in St. Johns. A currently underutilized employment area near the Willamette River may evolve into an area with more activity and a broader mix of land uses including housing. This area offers riverfront and view amenities that can attract development to further support the St. Johns commercial area.

The map author wishes to acknowledge Barry Manning, Lori Hill, Marguerite Feuersanger, Troy Doss, and Gary Odenthal.

Albury Street Map

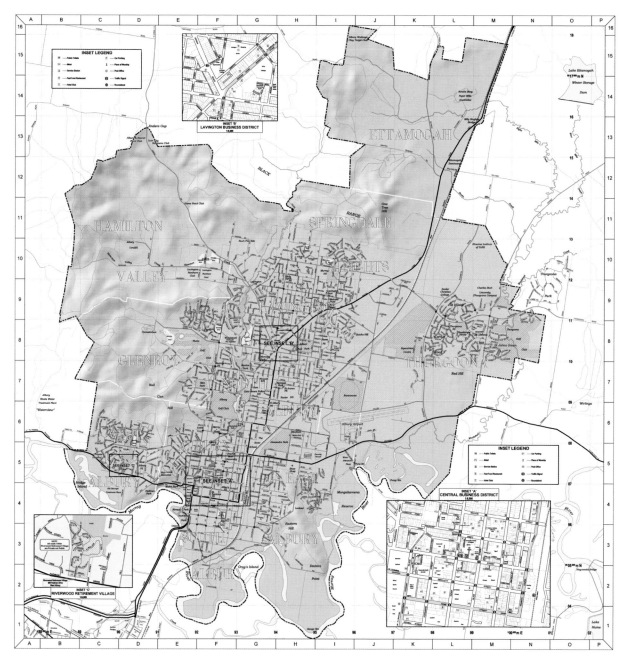

This map is one of a series of standard street directory types produced by AlburyCity, mainly for staff reference and customer service. They are also used as a base for other more specific thematic map productions. Printing is done in-house on demand, not in volume for general distribution. The maps are available to the public upon request and are produced in approximately A0 and A1 sizes, mainly in color. They are continually updated as the city grows. A0 maps feature larger scale insets and photographs of some of Albury's more significant heritage listed buildings. Some maps in this series are published from ArcMap and accessed by AlburyCity staff using ArcReader™.

AlburyCity (Local Government Authority)
Albury, New South Wales, Australia
By Russell Milton

Contact
Russell Milton
rmilton@alburycity.nsw.gov.au

Software
ArcInfo 8.3 and ArcReader

Hardware
Pentium 4 PC

Printer
HP Designjet 1055cm+

Government—State and Local

Region Vysocina Administrative Structure

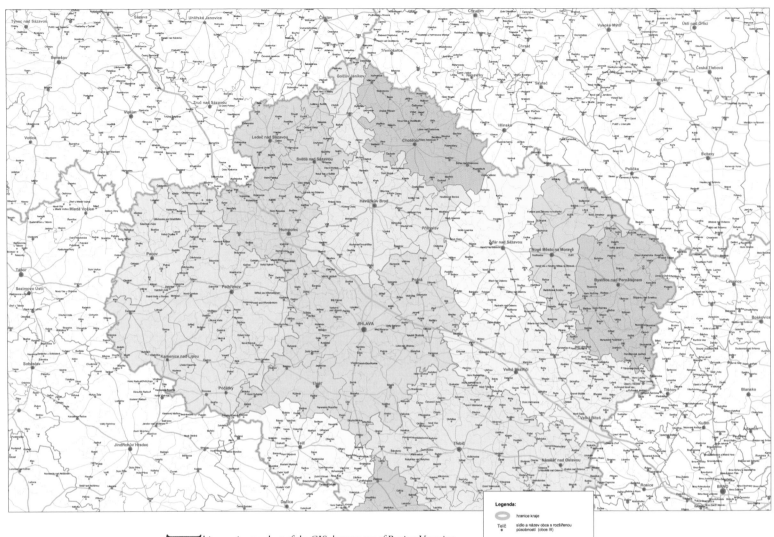

Region Vysocina

Jihlava, Czech Republic

By Jiri Hiess and Lubomir Juzl

Contact

Lubomir Juzl

gis@kr-vysocina.cz

Software

ArcIMS, ArcPress™, ArcView 3.3, and Adobe Illustrator

Hardware

Pentium 4

Printer

HP Designjet 1050c+

Data Source(s)

ArcCR500 and GIS Region Vysocina

Government—State and Local

This map is a product of the GIS department of Region Vysocina and was printed from 3,000 pieces in 1:200,000 scale. It is distributed free of charge within the region to end users and the public. This was the first attempt within the Czech Republic to describe the process of administrative transformation in Region Vysocina (7,000 square kilometers). With cooperation from other regions, this new administrative map was published on the Web via ArcIMS® (in August 2002).

Region Vysocina is an area of great changes in government. Today Vysocina has 729 municipal polygons. These small units are aggregated to larger polygons according to the branch of administration. New types of districts (205) are called "communities with enhanced force" and are shown on three types of map servers (ArcIMS, Minnesota MS, T-MapServer). Maps can be customized for particular needs and users can find a variety of maps including construction offices, registries, and tax offices. Data is regularly edited and updated. See Vysocina GIS online at http://gis.kr-vysocina.cz, which contains both a map gallery and projects useful for city managers, stakeholders, businesses, and citizens.

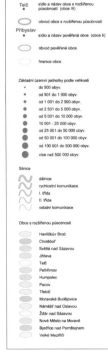

City of Toronto Building Construction Dates

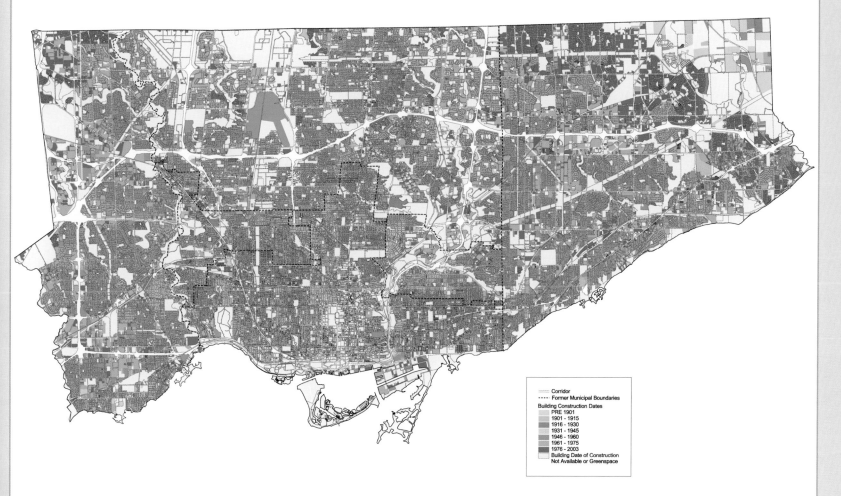

Legend:
- Corridor
- Former Municipal Boundaries

Building Construction Dates
- PRE 1901
- 1901 - 1915
- 1916 - 1930
- 1931 - 1945
- 1946 - 1960
- 1961 - 1975
- 1976 - 2003
- Building Date of Construction Not Available or Greenspace

Toronto, Canada, lies on the shore of Lake Ontario, the easternmost of the Great Lakes. Home to more than two million people, the city is the key to one of North America's most vibrant regions, the Greater Toronto Area (GTA). Four and one-half million Canadians live in the GTA, the cultural, entertainment, and financial capital of the nation. Toronto is the fifth largest municipal government in North America and is also the seat of the Ontario government. There are more than 466,700 assessed properties and 391,500 buildings within Toronto.

The City of Toronto Building Construction Dates map is a snapshot of growth trends in the city using year built information from assessment data. Building construction dates are derived from the Toronto Assessment Database and were extracted and linked to the city's cadastral fabric database through the assessment roll number. The city's parcel fabric was then dissolved on the year built and classified into a number of categories to highlight trends in the development and expansion of the city of Toronto.

City of Toronto
Toronto, Ontario, Canada
By Patricia Morphet

Contact
Patricia Morphet
pmorphe@toronto.ca

Software
ArcView 3.2a

Hardware
Dell Precision 530

Printer
Xerox Express-36 Inkjet

Data Source(s)
Municipal Property Assessment Corporation–Toronto Assessment Database and City of Toronto Works and Emergency Services–Technical Services, Survey and Mapping

Government—State and Local

Assessor's Map

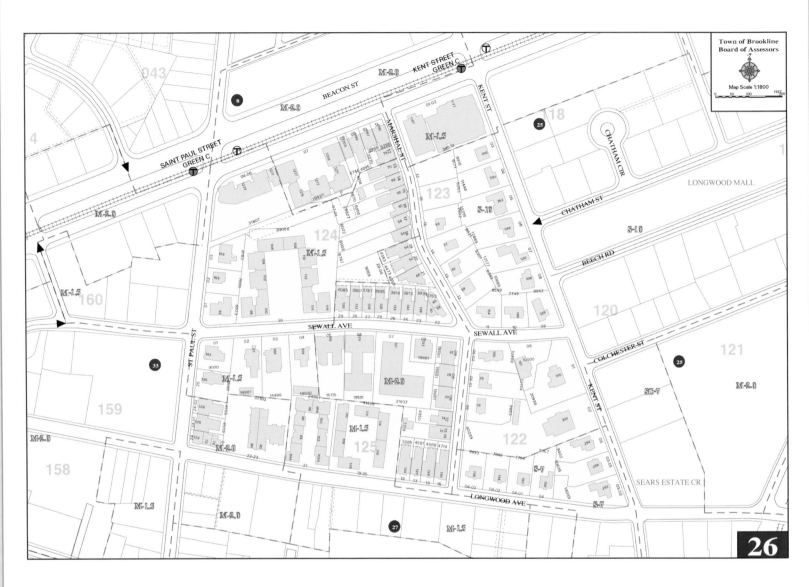

This map is a page in the Brookline Assessor's atlas book. It shows property boundaries with block, lot, and sublot numbers; lot sizes; easements; zoning districts; parks; streams; ponds; and subways. It is widely used in town departments and primarily by assessors for assessing property values. This is the first atlas book in Brookline produced using GIS. Previously all atlas book maps were drawn by hand, and updating was time-consuming and costly. With GIS, this book is updated when changes in the database are made, and it is less expensive to produce copies of the book for internal departments and the public.

The original map and atlas design is by Feng Yang, GISP. Members of the Brookline GIS group contributed to the data development. They are Alex Abbott, Parvaneh Kossari, Kate Lomen, Terry Meyer, Steve Petrecca, and Lynn Thorp. Valuable reviews and comments from the Brookline Assessor's Office were provided by Chief Assessor George Moody, Rachid Belhocine, Randy Kinkaid, and Linda MacDonald.

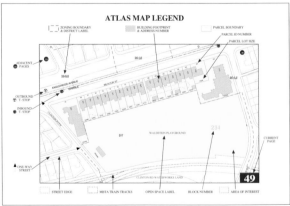

Town of Brookline

Brookline, Massachusetts, USA
By Feng Yang, GISP

Contact

Feng Yang
feng_yang@town.brookline.ma.us

Software

ArcInfo 8

Hardware

Sun Ultra 60

Printer

HP Color Laserjet 8550gn

Data Source(s)

Brookline GIS

Government—State and Local

Riverside County Blueprint for Tomorrow

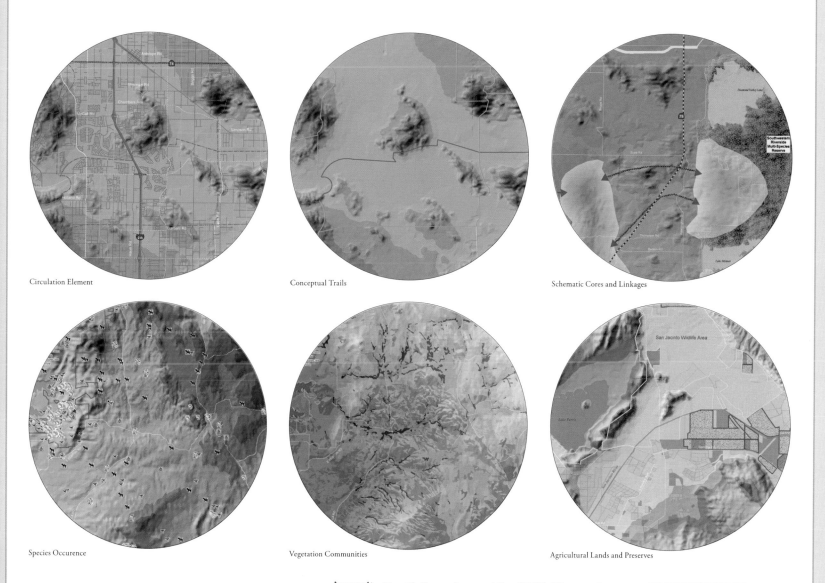

Circulation Element

Conceptual Trails

Schematic Cores and Linkages

Species Occurence

Vegetation Communities

Agricultural Lands and Preserves

The Riverside County Integrated Plan (RCIP), "Blueprint for Tomorrow," is a cooperative effort to address the exponential growth and urban sprawl currently underway in Riverside County. The county's population in 2000 was approximately 1.5 million people and is expected to increase nearly 400 percent during the next 40 years. Accommodating a population increase of this magnitude will involve urbanizing thousands of acres of undeveloped land.

Managing the planning process creatively in a manner that protects and increases a desirable quality of life while safeguarding the environment is a complex process and has been categorized into three basic RCIP components—General Plan/Land Use, Transportation Corridor Analysis, and Multiple Species Habitat Conservation Plan. This map document series depicts the organizational concept of the plan as well as details illustrating additional information crucial to the basic structure and implementation of the plan.

Riverside County Transportation and Land Management Agency GIS
Riverside, California, USA
By Sharon Baker–Stewart

Contact
Sharon Baker–Stewart
sbaker@co.riverside.ca.us

Software
ArcInfo, ArcGIS 3D Analyst, and Windows 2000

Printer
HP Designjet 1055cm+

Data Source(s)
California Department of Fish and Game, Riverside County Integrated Plan Consultants, U.S. Fish and Wildlife Service, and U.S. Geological Survey

Government—State and Local

Gwinnett County, Georgia, Countywide Data Integration for E911

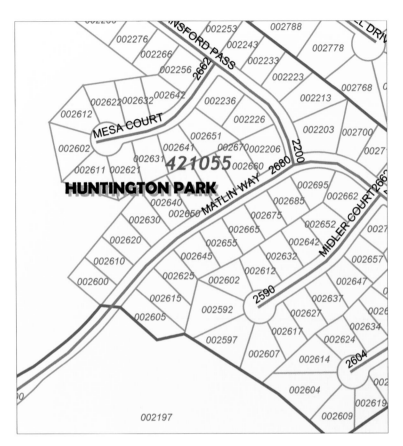

Gwinnett County, Georgia, is a rapidly growing community northeast of Atlanta. This map is part of a presentation depicting how, by using the county's enterprisewide GIS system and sources from several county and state agencies, Gwinnett County maintains a street centerline network with street, address, and other location information that meets the standards necessary to support the computer-aided dispatching of police, fire, and emergency services.

Using scanned images of recorded plats, text records of real property from the tax assessor, raster aerial photography, and vector features showing rights-of-way in the cadastral layer, county staff developed and maintains a data set of more than 28,000 street segments in an enterprise ArcSDE geodatabase. Extracted, the data passes through several processes in Tiburon Inc.'s Geographic Conversion Toolkit to prepare a geofile for the CADTi computer-aided dispatch system.

Gwinnett County

Lawrenceville, Georgia, USA

By Sharon C. Stevenson, GISP

Contact

Sharon Stevenson

sharon.stevenson@gwinnettcounty.com

Software

ArcMap 8.2, ArcSDE, and Windows 2000

Hardware

Gateway Professional S PC

Printer

HP Designjet 1050c

Data Source(s)

Georgia Superior Court Clerks'
Cooperative Authority and Gwinnett County

Government—State and Local

West Nile Virus in Illinois—2001 and 2002

Social and environmental factors that may contribute to the risk of WNV infection including the Mosquito Abatement Districts and the physical characteristics of the landscape

Additional factors to consider including a comparison of dead bird locations in 2001 and human cases in 2002 and SLE cases in 1975 relative to WNV cases in 2002

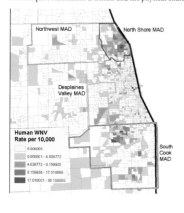

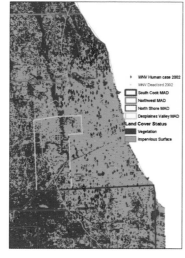

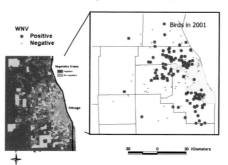

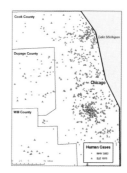

Progression from May 5 to October 10

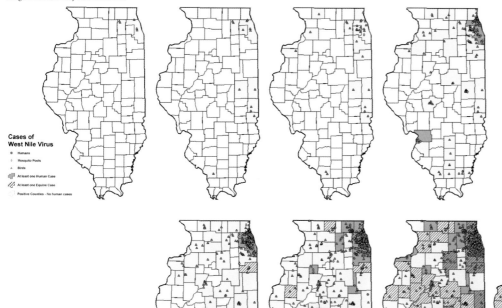

West Nile virus has spread across the nation since it was first observed in the New York City area in 1999. In 2001, West Nile virus was found in dead birds in only a few Illinois counties. During the following year, Illinois topped all states in the number of human cases of West Nile virus, logging more than 877 cases and 63 deaths. The counties of Cook and Dupage, in the Chicago region, accounted for approximately 80 percent of those cases. In the course of the 2002 season, it spread through the bird, mosquito, horse, and human populations, with cases across the state.

Spatial analysis and sequential mapping are useful and powerful tools to explore the patterns and processes related to the introduction and spread of a new disease. The College of Veterinary Medicine GIS and Spatial Analysis Lab and the Spatial Epidemiology Lab helped the Illinois Department of Public Health and Illinois Department of Agriculture with mapping and analysis of the West Nile virus in Illinois during the recent outbreak. The images demonstrate the various mapping approaches taken and a summary analysis of the human case clusters that occurred in the Chicago region.

University of Illinois
Urbana, Illinois, USA
By Julie Clennon, Uriel Kitron,
Aaron Lippold, Thomas J. McTighe,
David Norris, and Marilyn O. Ruiz

Contact
Thomas J. McTighe
mctighe@uiuc.edu

Software
ArcGIS 8, ArcIMS 4 for Red Hat Linux,
ArcSDE 8 for Red Hat Linux,
CorelDRAW 11, Red Hat Linux 7.3,
and Windows 2000

Printer
HP Designjet 500

Data Source(s)
College of Veterinary Medicine, Illinois
Department of Agriculture, and Illinois
Department of Public Health

Health and Human Services

West Texas Atlas of Rural and Community Health

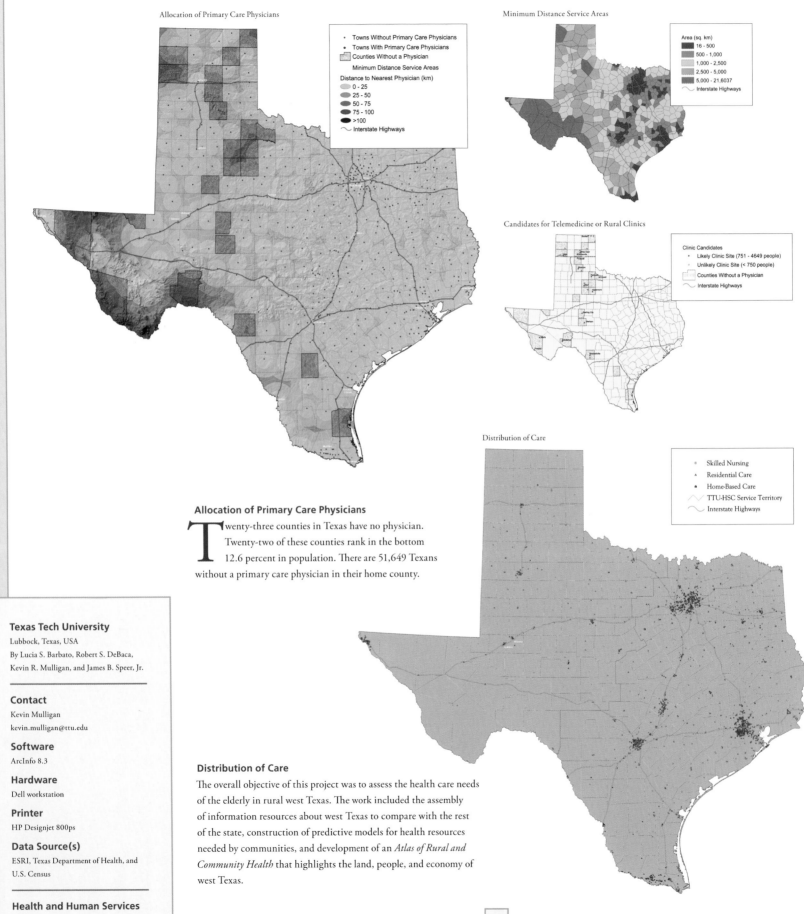

Allocation of Primary Care Physicians

Minimum Distance Service Areas

Candidates for Telemedicine or Rural Clinics

Distribution of Care

Allocation of Primary Care Physicians

Twenty-three counties in Texas have no physician. Twenty-two of these counties rank in the bottom 12.6 percent in population. There are 51,649 Texans without a primary care physician in their home county.

Texas Tech University

Lubbock, Texas, USA

By Lucia S. Barbato, Robert S. DeBaca, Kevin R. Mulligan, and James B. Speer, Jr.

Contact

Kevin Mulligan

kevin.mulligan@ttu.edu

Software

ArcInfo 8.3

Hardware

Dell workstation

Printer

HP Designjet 800ps

Data Source(s)

ESRI, Texas Department of Health, and U.S. Census

Health and Human Services

Distribution of Care

The overall objective of this project was to assess the health care needs of the elderly in rural west Texas. The work included the assembly of information resources about west Texas to compare with the rest of the state, construction of predictive models for health resources needed by communities, and development of an *Atlas of Rural and Community Health* that highlights the land, people, and economy of west Texas.

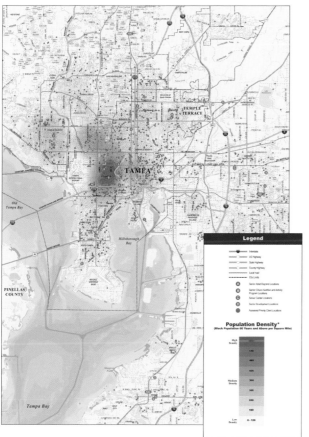

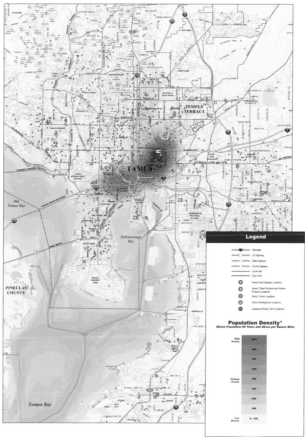

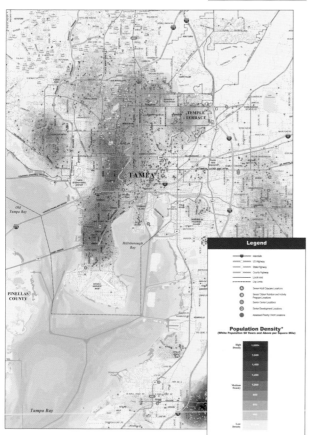

During the summer of 2002, Hillsborough County's Aging Services Department received funding to expand its service center locations. Having only the addresses of program enrollees, the Aging Services Department contacted the Real Estate Department's GIS Section to spatially locate clients so that locations could be chosen to minimize the driving distances and travel time to the new centers.

The Aging Services Department also recognized the need to provide services to people who would be enrolling in its program in the coming years. The GIS Section used the 2000 U.S. Census Bureau data to determine the locations of potential clients, identifying those areas where people aged 60 and over were living. Density maps were created using ArcGIS Spatial Analyst. Using this data, the GIS Section was able to determine several available parcels that would maximize a new center's usefulness to program enrollees. By locating the expansion centers on vacant, county-owned land when possible, the department was further able to lower costs.

Hillsborough County
Tampa, Florida, USA
By Dan Hardy

Contact
Rich Cvarak
cvarakr@hillsboroughcounty.org

Dan Hardy
hardyd@hillsboroughcounty.org

Software
ArcGIS Spatial Analyst, ArcView 3.2,
Adobe Illustrator, MAPublisher, and
Windows XP

Hardware
Compaq SP750 Workstation

Printer
HP Designjet 5000

Data Source(s)
U.S. Census Bureau and various local
sources

Health and Human Services

The Plague of Disease and Poverty in Sub-Saharan Africa

The Seven Leading Causes of Death in Sub-Saharan Africa

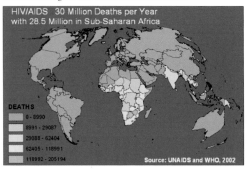

HIV/AIDS 30 Million Deaths per Year with 28.5 Million in Sub-Saharan Africa

DEATHS
- 0 - 8990
- 8991 - 29087
- 29088 - 62404
- 62405 - 118991
- 118992 - 205194

Source: UNAIDS and WHO, 2002

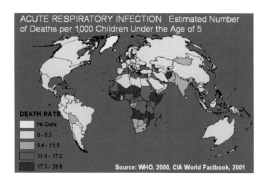

ACUTE RESPIRATORY INFECTION Estimated Number of Deaths per 1,000 Children Under the Age of 5

DEATH RATE
- No Data
- 0 - 5.3
- 5.4 - 11.5
- 11.6 - 17.2
- 17.3 - 28.8

Source: WHO, 2000, CIA World Factbook, 2001

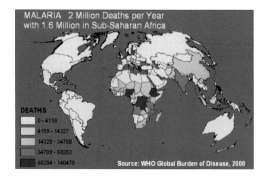

MALARIA 2 Million Deaths per Year with 1.6 Million in Sub-Saharan Africa

DEATHS
- 0 - 4158
- 4159 - 14327
- 14328 - 34708
- 34709 - 60283
- 60284 - 140478

Source: WHO Global Burden of Disease, 2000

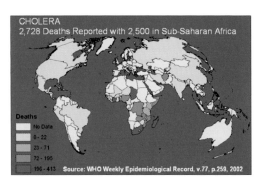

CHOLERA 2,728 Deaths Reported with 2,500 in Sub-Saharan Africa

Deaths
- No Data
- 0 - 22
- 23 - 71
- 72 - 195
- 196 - 413

Source: WHO Weekly Epidemiological Record, v.77, p.259, 2002

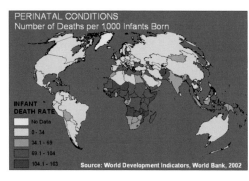

PERINATAL CONDITIONS Number of Deaths per 1,000 Infants Born

INFANT DEATH RATE
- No Data
- 0 - 34
- 34.1 - 69
- 69.1 - 104
- 104.1 - 163

Source: World Development Indicators, World Bank, 2002

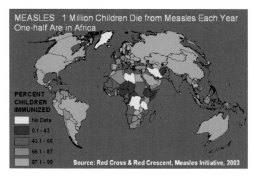

MEASLES 1 Million Children Die from Measles Each Year One-half Are in Africa

PERCENT CHILDREN IMMUNIZED
- No Data
- 0.1 - 43
- 43.1 - 66
- 66.1 - 87
- 87.1 - 99

Source: Red Cross & Red Crescent, Measles Initiative, 2003

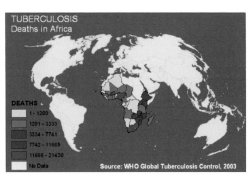

TUBERCULOSIS Deaths in Africa

DEATHS
- 1 - 1200
- 1201 - 3333
- 3334 - 7741
- 7742 - 11665
- 11666 - 21430
- No Data

Source: WHO Global Tuberculosis Control, 2003

Sub-Saharan Africa suffers high death rates from curable diseases because of the regional problem of severe poverty. Six of the top seven causes of death in this area are preventable. HIV/AIDS is the one cause that is not yet curable; 95 percent of HIV deaths occur in sub-Saharan Africa. The remaining top six causes of death—acute respiratory infection, malaria, diarrheal disease, prenatal conditions, measles, and tuberculosis—are all curable or preventable but remain a problem in sub-Saharan Africa because of severe poverty.

The leading indicators of poverty are low income, poor access to improved water, illiteracy, poor medical services, low life expectancy, and malnourishment. These maps express these indicators and show how sub-Saharan Africa is trapped in severe poverty and needless death. This visualization will hopefully prompt action to help break this desperate cycle.

Global Infosci

Post Falls, Idaho, USA

By Ed DeYoung

Contact

Ed DeYoung

ed@globalinfosci.com

Software

ArcView 8.2 and Windows 2000

Printer

HP large format printer

Data Source(s)

Joint United Nations Programme on HIV/AIDS, Red Crescent, Red Cross, United Nations, World Bank, and World Health Organization

Health and Human Services

Main Causes of Poverty in the World

ACCESS TO IMPROVED WATER

% People with Improved Water
- 24 - 47
- 48 - 65
- 66 - 81
- 82 - 90
- 91 - 100
- No Data

Source: World Development Indicators, World Bank, 2003

ADULT LITERACY RATE

Percent Literate
- 0 - 45
- 46.1 - 68
- 69 - 86
- 87 - 100
- No Data

Source: CIA World Factbook, 2000

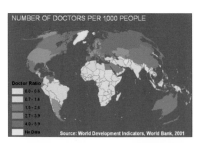

NUMBER OF DOCTORS PER 1,000 PEOPLE

Doctor Ratio
- 0.0 - 0.6
- 0.7 - 1.4
- 1.5 - 2.6
- 2.7 - 3.9
- 4.0 - 5.9
- No Data

Source: World Development Indicators, World Bank, 2001

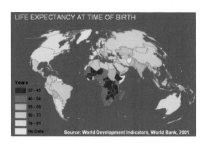

PURCHASING POWER PARITY

1998 DOLLARS
- $0 - $168
- $170 - $453
- $454 - $928
- $929 - $1844
- $1845 - $3950
- No Data

Source: World Development Indicators, World Bank, 2001

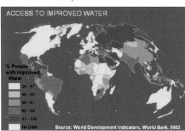

LIFE EXPECTANCY AT TIME OF BIRTH

Years
- 37 - 45
- 46 - 54
- 55 - 65
- 66 - 73
- 74 - 81
- No Data

Source: World Development Indicators, World Bank, 2001

MALNOURISHMENT

% Children Underweight
- 1.0 - 9.0
- 9.1 - 18.0
- 18.1 - 27.0
- 27.1 - 34.0
- 34.1 - 48.0
- No Data

Source: World Development Indicators, World Bank, 2003

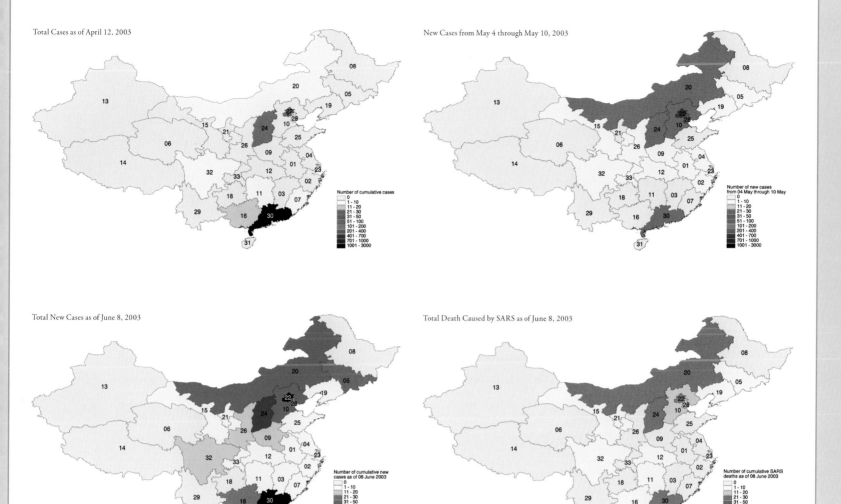

Total Cases as of April 12, 2003

New Cases from May 4 through May 10, 2003

Total New Cases as of June 8, 2003

Total Death Caused by SARS as of June 8, 2003

The Severe Acute Respiratory Syndrome (SARS) outbreak was first observed in early 2003. Medical communities believed the outbreak originated in southern China. In March 2003 in Vancouver and Toronto, Canada, cases with similar symptoms were reported by individuals who had recently traveled to China. It became a significant concern throughout the world when medical officials observed that people developed symptoms after a visit to China, resulting in local outbreaks. Fear spread throughout the world because not much was known about the disease. There was no known cause, no sure means of distinguishing the disease, no known successful treatment, and a lack of information on its incubation period.

The creator of the map hoped to find some clues about the spread of the disease by interpreting statistical data from medical communities and displaying the data geographically. The data, provided by the Chinese Ministry of Health, was collected from the World Health Organization Web site. The purpose of the map was to develop a statistical analysis of the data provided. The map shows only a preliminary review of data, visually identifying patterns and trends of movements. The accuracy of the data is unknown because reporting methods are not clear.

National Geospatial-Intelligence Agency
Washington, D.C., USA
By Yoon Cho

Contact
Yoon Cho
choys@nga.mil

Software
ArcView 3.3 and Microsoft Excel

Hardware
DataBlade running on UNIX

Printer
HP

Data Source(s)
National Geospatial-Intelligence Agency and World Health Organization

Health and Human Services

Mapping Health Care Delivery for Children Using Primary Care Service Areas

U.S. Primary Care Service Areas

PCSAs are randomly colored.
N = 6,542 (v2.1)

State Boundaries
Census Unassigned ZCTAs
Inland Waters

Children Under 18 Per Clinically Active Pediatrician

Other Areas
State Boundaries
Census Unassigned ZCTAs
Inland Waters

Fewer than 1000 (86)
1000 - 2000 (137)
2000 - 4000 (151)
4000 or more (72)
No clinically active pediatrician (332)

Median Household Income

Other Areas
State Boundaries
Census Unassigned ZCTAs
Inland Waters

$46,002 - $146,755 (161)
$37,916 - $46,000 (144)
$33,178 - $37,910 (131)
$29,186 - $33,176 (132)
$9,783 - $29,186 (210)

Percent of Clinically Active Pediatricians Who Are Female

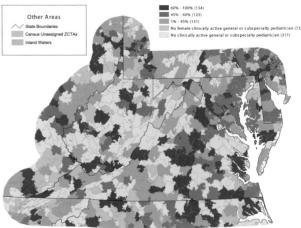

Other Areas
State Boundaries
Census Unassigned ZCTAs
Inland Waters

60% - 100% (134)
45% - 60% (123)
1% - 45% (131)
No female clinically active general or subspecialty pediatrician (73)
No clinically active general or subspecialty pediatrician (317)

As part of an effort to develop a national database of primary care (i.e., office-based or outpatient care) resources and health services, the Center for the Evaluative Clinical Sciences at Dartmouth Medical School has developed small, standardized geographic markets that reflect primary care utilization. These areas are called Primary Care Service Areas (PCSAs). PCSAs are defined by assigning U.S. Census 2000 ZIP Code Tabulation Areas to primary care providers based on the primary care utilization of Medicare beneficiaries.

For online information on the PCSA project, visit http://datawarehouse.hrsa.gov/pcsa.htm.

The uneven geographic distribution of health care services remains one of the important challenges facing U.S. health policy makers. In the area of children's health, the American Academy of Pediatrics (AAP) has partnered with the Center for the Evaluative Clinical Sciences at Dartmouth Medical School to develop an interactive Web-based database that will be used to improve the provision of pediatric care by supplying information on the demographics, socioeconomic status, and geographic distribution of children and providers of pediatric care.

Using primary care service areas as the geographic unit, the project will identify patterns in demographics, physician supply, geographic distributions of primary care pediatricians and pediatric subspecialists, and locations of federally funded health clinics. In a first release of the data, online U.S. and state AAP maps with associated data tables are viewable at http://www.aap.org/mapping/.

Dartmouth College

Hanover, New Hampshire, USA
By David C. Goodman, Nancy Marth, and James Poage

Contact

Nancy Marth
nancy.marth@dartmouth.edu

Software

ArcInfo 8.3, ArcView 3.3, and Windows 2000 Pro

Hardware

Dell Precision 330

Printer

HP Designjet 800ps

Data Source(s)

Medicare, U.S. Census, and physician resource data

Health and Human Services

Developing North American Soil Properties for Climate and Hydrology Applications—Mexico

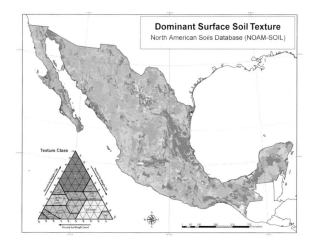

Dominant Surface Soil Texture
North American Soils Database (NOAM-SOIL)

Texture Class

Soil Polygon-Pedon Match Score

Match Score
1 - 5
6 - 10
11 - 20
Spatially coincident pedons

Distance to Pedon Component 1

Distance
< 50 km
50 - 100 km
100 - 250 km
250 - 500 km
500 - 1000 km
> 1000 km
Spatially coincident pedons

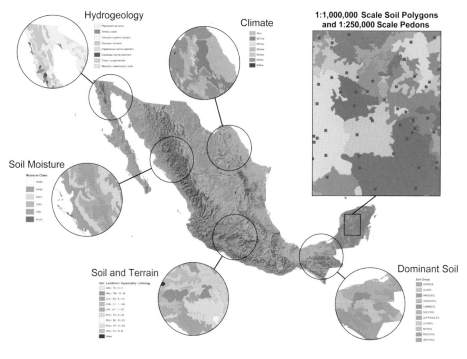

Hydrogeology

Climate

1:1,000,000 Scale Soil Polygons and 1:250,000 Scale Pedons

Soil Moisture

Moisture Class

Soil and Terrain

Dominant Soil

Soil Group

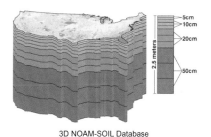

3D NOAM-SOIL Database

The North American Soil Properties (NOAM-SOIL) project goal is to create a three-dimensional soil properties geographic database suitable for use with climate and hydrologic models for North America. An automated GIS method patterned after manual procedures used by pedologists to link existing soil physical property measurements with soil map polygons was developed; 3,037 georeferenced pedon measurements were evaluated for linkage to the first, second, and third soil components for 8,260 polygons of the 1:1,000,000-scale Soil Map of Mexico. The linkage procedure relied on distance from the polygon and a goodness-of-match score that compared the polygon soil component and the soil pedon taxonomy, physical/chemical phase, and landscape location attributes.

Pedons that were spatially coincident with a soil polygon and matched at the soil type (suborder) level were assigned as the "best" pedon. The number of matches of soil phase data resolved conflicts.

Remaining soil polygons were processed for each of the three soil components and each of five complementary spatial databases. For each polygon, the pedon nearest to a polygon node that matched the soil component's soil type and the value of the complementary database was added to the polygon's list of matching pedons.

The best pedon was assigned to each soil polygon component based on the number of complementary database matches, the number of phase matches, and the distance between the pedon and the polygon edge.

A dominant soil surface texture map with metadata maps for the Mexican portion of North America is among preliminary NOAM-SOIL products. NOAM-SOIL is supported by the National Oceanic and Atmospheric Administration–Global Energy Water-Cycle Experiment Continental-Scale International Project No. GC98-300. The authors wish to acknowledge the contributions of Instituto Nacional de Estadistica, Geografia e Informatica (INEGI) and Inventorio Nacional de Suelos-DGRyCS-Secretaria de Medio Ambiente y Recursos Naturales–Mexico.

U.S. Department of Agriculture, Agricultural Research Service, National Soil Tilth Laboratory
Ames, Iowa, USA

U.S. Department of Agriculture, Natural Resources and Conservation Service, National Soil Survey Center
Lincoln, Nebraska, USA
By David E. James and Sharon W. Waltman

Contact
Sharon Waltman
sharon.waltman@usda.gov

Software
ArcInfo 8.2 Workstation and Windows 2000

Hardware
Dell Precision Workstation 530

Printer
HP Designjet 3500cp

Data Source(s)
Global and National Soils and Terrain Digital Database; National Commission for the Knowledge and Use of Biodiversity, Mexico; Secretariat of the Environment, Natural Resources, and Fisheries; Soil Maps of Mexico; and United Nations Food and Agriculture Organization

Natural Resources—Agriculture

U.S. Department of Agriculture, National Agricultural Statistics Service—Geospatial Products and Data

**Corn for 2002
Planted Acres by County**

Acres
Not Estimated
< 5,000
5,000 - 9,999
10,000 - 24,999
25,000 - 49,999
50,000 - 99,999
100,000 - 149,999
150,000 +

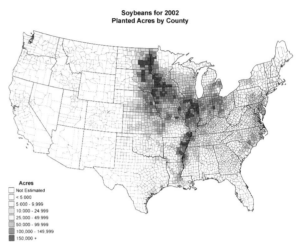

**Soybeans for 2002
Planted Acres by County**

Acres
Not Estimated
< 5,000
5,000 - 9,999
10,000 - 24,999
25,000 - 49,999
50,000 - 99,999
100,000 - 149,999
150,000 +

U.S. Department of Agriculture, National Agricultural Statistics Service

Fairfax, Virginia, USA

By T. Lloyd, D. Norton, L. Plowman, and G. Wade

Contact

Lee Plowman

lee_plowman@nass.usda.gov

Software

ArcGIS, ArcInfo, ArcView, Atlas GIS™, UNIX, and Windows 2000

Hardware

PC and UNIX workstation

Printer

HP Designjet 5000ps

Data Source(s)

1997 Census of Agriculture, USDA/NASS surveys, satellite imagery, and topographic vector and raster data

Natural Resources—Agriculture

Land Use Stratification Maps

Land use stratification maps are produced by the U.S. Department of Agriculture, National Agricultural Statistics Service (USDA/NASS) and depict agricultural land use within a state for statistical sampling purposes. Image analysts at NASS use satellite imagery, topographic data, and agriculture surveys as data sources for stratifying the land. The stratification process divides land into broad categories—cultivated land, natural vegetation, urban areas, nonagricultural land, and large bodies of water. The cultivated land is further subdivided and categorized by the intensity of agricultural activity, which is represented as a percentage of cultivation. Stratification maps and shapefiles are accessible at http://www.nass.usda.gov/research/stratafront2b.htm.

Annual County-Level Crop Acreage and Yield Maps

The NASS County Estimates program annually collects crop acreage and yield data through cooperative agreements with each state by mailing surveys to a large sample of farm operators. The county crop acreage and yield estimates are assembled starting with the state estimate and working back to the county-level estimates. The county-level estimates are published on the Internet for each USDA state statistical office. The county-level estimates of crop acreage and yield are used to develop choropleth maps for NASS. The maps are available for public access on the USDA/NASS Web site at http://www.usda.gov/nass/aggraphs/cropmap.htm.

**Stratification of Iowa
1989**

Land Use Strata
> 75 % Cultivated
25 - 75 % Cultivated
< 25 % Cultivated
Agri-Urban > 100 Homes Per Sq. Mi.
Commercial/Dense Urban > 100 Homes Per Sq. Mi.
Non-Agriculture
Water

**Stratification of Tennessee
2001**

Land Use Strata
> 50 % Cultivated
15 - 50 % Cultivated
< 15 % Cultivated
Agri-Urban > 100 Homes Per Sq. Mi.
Commercial/Dense Urban > 100 Homes Per Sq. Mi.
Non-Agriculture
Water

U.S. Department of Agriculture, National Agricultural Statistics Service—Geospatial Products and Data

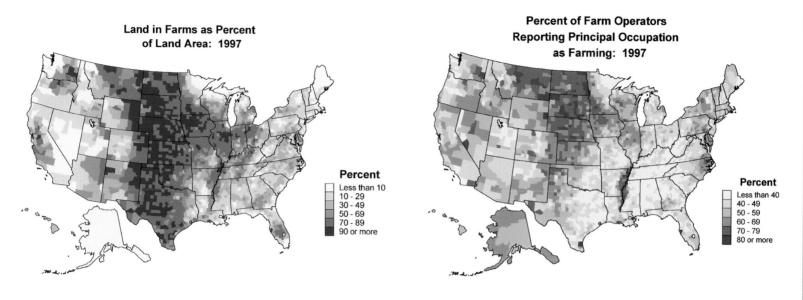

Land in Farms as Percent of Land Area: 1997

Percent
- Less than 10
- 10 - 29
- 30 - 49
- 50 - 69
- 70 - 89
- 90 or more

Percent of Farm Operators Reporting Principal Occupation as Farming: 1997

Percent
- Less than 40
- 40 - 49
- 50 - 59
- 60 - 69
- 70 - 79
- 80 or more

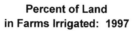

Percent of Land in Farms Irrigated: 1997

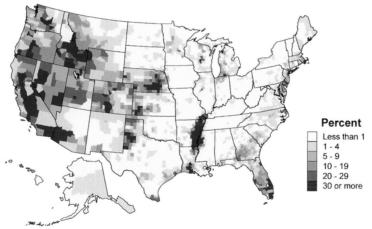

Percent
- Less than 1
- 1 - 4
- 5 - 9
- 10 - 19
- 20 - 29
- 30 or more

Agricultural Atlas of the United States

Choropleth dot distribution maps from the 1997 Census of Agriculture, Agricultural Atlas of the United States illustrate various aspects of the nation's agriculture. The maps portray major topics from the U.S. 1997 Census of Agriculture including crops, livestock, farm and operator characteristics, land use, agricultural chemicals and equipment, and economic themes. Atlas maps of the U.S. 2002 Census of Agriculture will be available to the public on the USDA/NASS Web site in 2004. The 1997 agricultural atlas maps are at http://www.nass.usda.gov/census/census97/atlas97/index.htm.

U.S. Department of Agriculture, National Agricultural Statistics Service

Fairfax, Virginia, USA

By T. Lloyd, D. Norton, L. Plowman, and G. Wade

Contact

Lee Plowman

lee_plowman@nass.usda.gov

Software

ArcGIS, ArcInfo, ArcView, Atlas GIS, UNIX, and Windows 2000

Hardware

PC and UNIX workstation

Printer

HP Designjet 5000ps

Data Source(s)

1997 Census of Agriculture, USDA/NASS surveys, satellite imagery, and topographic vector and raster data

---

Natural Resources—Agriculture

Agriculture Maps of South Africa

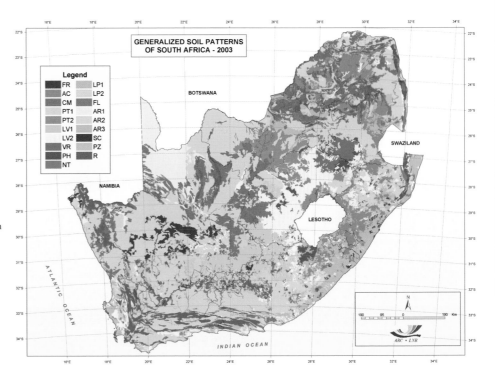

Generalized Soil Patterns of South Africa 2003

The land type survey is a reconnaissance survey that followed an inventory rather than a fixed legend approach. The soils were subsequently reorganized based on pedogenesis and land use capability and grouped into 19 generalized soil patterns to produce a map with a scale of 1:1,000,000. In this process, wherever possible, polygons smaller than 0.25 square centimeters (3,900 hectares) were incorporated into the largest bordering polygon with similar characteristics. International soil classification standards were also followed where possible.

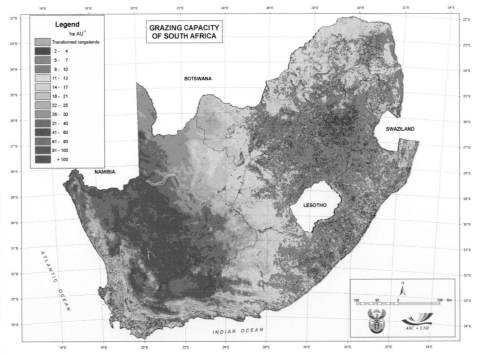

Agriculture Research Council– Institute for Soil, Climate, and Water

Pretoria, South Africa

By T. Germishuyse and M. Samadi (Generalized Soil Patterns)

T. Newby, D.J. Pretorius, and H.J.C. Smith (Grazing Capacity)

Contact

Marjan van der Walt

marjan@iscw.agric.za

Software

ArcGIS Spatial Analyst, ArcInfo 8.3, and Microsoft Access 2000

Hardware

Pentium 4

Printer

HP Color Laserjet 4600n

Data Source(s)

Land cover and land type data, normalized difference vegetation index, and tree density data

Natural Resources—Agriculture

Grazing Capacity of South Africa

This map refined a draft grazing capacity map by correlating the maximum normalized difference vegetation index (NDVI) image of 200 with animal unit values from earlier grazing capacity maps. Refinements took the form of incorporating land cover and tree density data. The integration of tree density data was based on a number of underlying assumptions: that a linear relationship exists between tree density and biomass production as measured by NDVI, 0–10 percent tree density can be regarded as having no influence on biomass production, at 75 percent tree density there would be very low to no biomass production, and the influence would not be the same for all tree species. The long-term NDVI map with tree density integrated was converted into grazing capacity values. Where overlapping occurs with transformed rangeland, the grazing capacity values were masked with land cover classes (cultivated land, forest plantations, mines and quarries, urban and water bodies). Adjustment of NDVI values was not attempted.

Agriculture Maps of South Africa

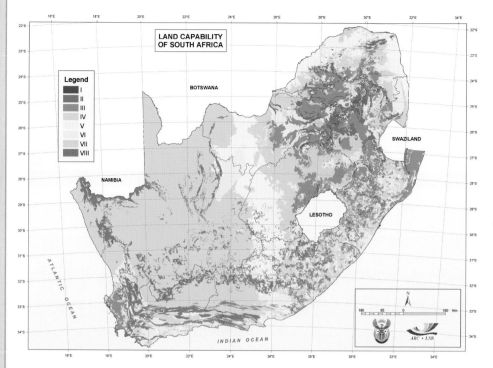

Land Capability of South Africa

The classic eight-class land capability system was adapted for use with GIS in South Africa taking data availability into account. Land capability classes are interpretive groupings of land units with similar potentials and continuing limitations or hazards. Land capability is a more general term than land suitability and more conservation oriented. It involves consideration of the risks of land damage from erosion and other causes and the difficulties in land use owing to physical land characteristics including climate. Social and economic variables are not considered.

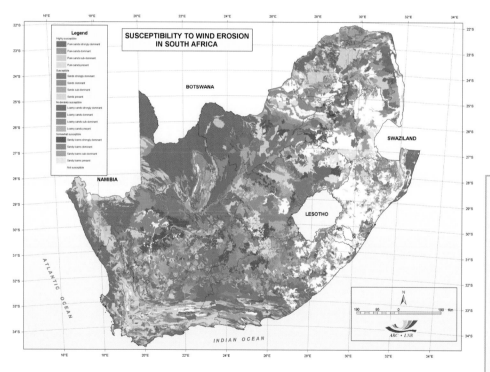

Susceptibility to Wind Erosion in South Africa

In the western and northwestern cultivation districts of the Western Cape, North West and Free State Provinces, wind erosion is an acknowledged problem. This is because of the prevalence of sandy soils, high winds, cultivation practices, and low rainfall resulting in low plant biomass production and low soil organic material. Wind erosion is the loss of fine materials (fine silt and clay) from topsoil in the form of dust. By losing fine material, the soil loses much of its ability to provide plants with water and nutrients. The main factors determining susceptibility to wind erosion are particle size distribution of the topsoil, wind speed, topography, soil cover, soil water content, and aggregation of soil particles. In producing this coverage, the emphasis was on particle size distribution as the prime permanent factor rendering land susceptible to wind erosion. It follows that the higher the sand (especially fine and very fine sand) fractions, the more susceptible the soil is to wind erosion. Silt could not be used as a parameter because of lack of data.

Agriculture Research Council–Institute for Soil, Climate, and Water

Pretoria, South Africa

By J. Malherbe, J.L. Schoeman, and M. van der Walt (Land Capability)

J.L. Schoeman, M. van der Walt, and Land Type Survey staff (Wind Erosion)

Contact

Marjan van der Walt

marjan@iscw.agric.za

Software

ArcGIS Spatial Analyst, ArcInfo 8.3, and Microsoft Access 2000

Hardware

Pentium 4

Printer

HP Color Laserjet 4600n

Data Source(s)

AgroMet Databank and land type data

Natural Resources—Agriculture

Investigating Brown Rot in Stone Fruit Using ArcIMS

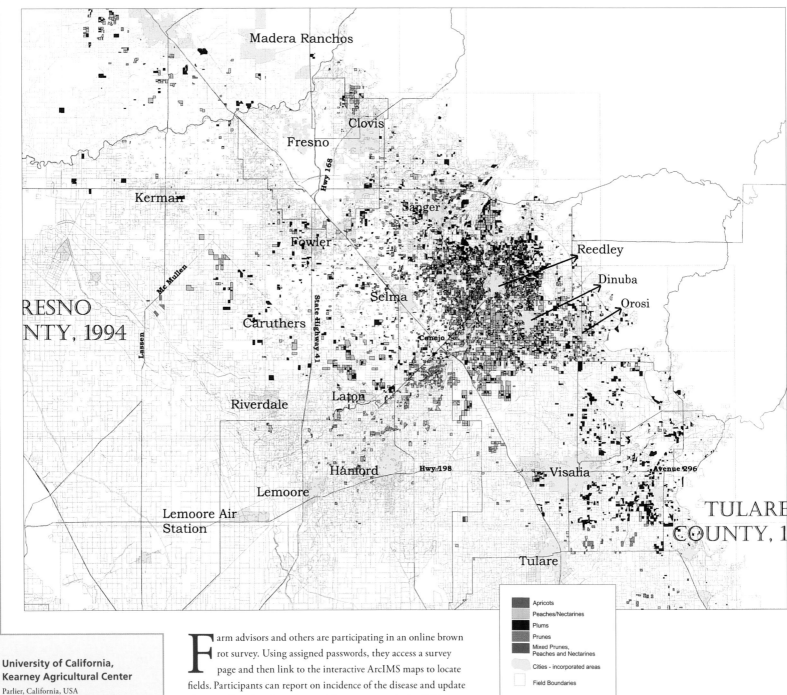

Farm advisors and others are participating in an online brown rot survey. Using assigned passwords, they access a survey page and then link to the interactive ArcIMS maps to locate fields. Participants can report on incidence of the disease and update the stone fruit distribution maps. They can query the ArcIMS map interface by township or section to locate areas of investigated stone fruit orchards. The Identify tool is used to get an identification number, which is entered into the survey form to provide the spatial reference to identify brown rot and changes to the overall map.

University of California, Kearney Agricultural Center

Parlier, California, USA

By Kris Lynn–Patterson and Themis Micheilides

Contact

Kris Lynn–Patterson

krislynn@uckac.edu

Software

ArcIMS, ArcView 3.3, and Windows 2000

Data Source(s)

California Department of Water Resources and California Spatial Information Library

Natural Resources—Agriculture

Dinan Bay Forest Development Plan Page 1

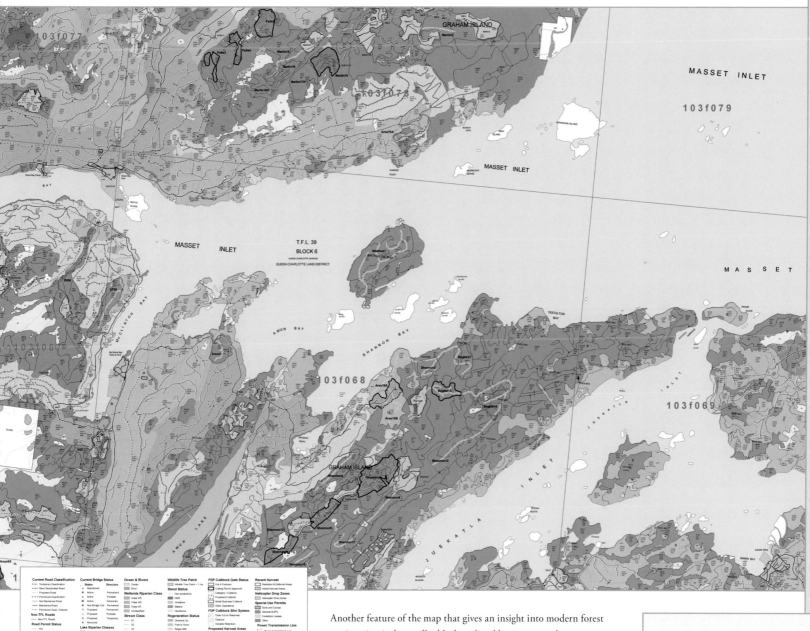

Another feature of the map that gives an insight into modern forest engineering is the smaller black outlined harvest areas that are developed from the red proposed areas.

This map formulates a part of the annual forest development plan (FDP) for Weyerhaeuser Company Ltd., Queen Charlotte Islands Timberlands Unit. Represented are projected logging areas and the existing regeneration status of the forest. On inspection, the transition from the old-style large clear-cut to a much smaller variable retention style of harvest area, which mimics more closely natural forest openings and disturbances, can be seen.

The map atlas is one of 51 pages (36- by 48-inch sheets) in a map atlas, which accompanies an FDP. The FDP is submitted to the British Columbia Ministry of Forests for approval of harvesting proposals under the forest practices code.

The author would like to acknowledge the contribution of Dave Trim, unit planner of the Queen Charlotte Islands Timberlands Unit, Weyerhaeuser Company Ltd.

Weyerhaeuser Company Ltd.
Queen Charlotte Islands,
British Columbia, Canada
By Nicholas Macdonald, CD

Contact
Nicholas Macdonald, CD
nick.macdonald@lpcorp.com

Software
ArcMap

Printer
HP Designjet 2500cp

Data Source(s)
Aerial photos, field notes, and other available coverages

Natural Resources—Forestry

Land Cover of North America

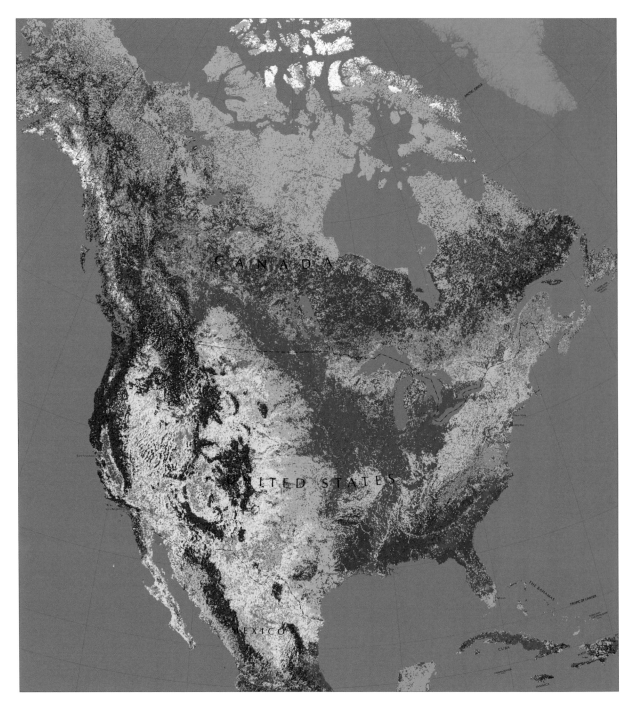

United Nations Environment Programme
Sioux Falls, South Dakota, USA
By Mark A. Ernste, Chandra Giri, and
John Hutchinson

Contact
Mark Ernste
ernste@usgs.gov

Software
ArcInfo Workstation, Adobe Illustrator,
and Adobe Photoshop

Hardware
Apple Macintosh G4 and SGI UNIX Server

Printer
Harris five-color offset press

Data Source(s)
GTOPO30 global topographic data,
National Atlas of the United States, and
SPOT Vegetation satellite data

Natural Resources—Forestry

The availability and accessibility of regional and global land cover data sets play an important role in many global change studies including climate change, biodiversity assessment, ecosystem assessment, and increasingly sophisticated Earth system models. The importance of such science-based information is reflected in a growing number of international, continental, and national projects and programs.

The land cover characterization database of North America for the year 2000 was prepared jointly by the U.S. Geological Survey's Earth Resources Observation System Data Center and the Canada Center for Remote Sensing using SPOT Vegetation satellite data. This is a regional component of the Global Land Cover 2000 (GLC-2000) project. A detailed description of the methodology used and data sets produced is at http://landcover.usgs.gov/glcc/index.asp. The Joint Research Center of the European Commission in collaboration with more than 30 partner institutions around the world is implementing the GLC-2000 project (www.gvm.sai.jrc.it/glc2000/defaultGLC2000.html). The main objective of the project is to prepare a consistent and reliable land cover database of the world for the year 2000. Map compilation and printing are provided by the United Nations Environment Programme, Sioux Falls, South Dakota. For a free copy of this poster, contact Jane S. Smith at jssmith@usgs.gov.

A New Method to Represent Temporal Data Sets

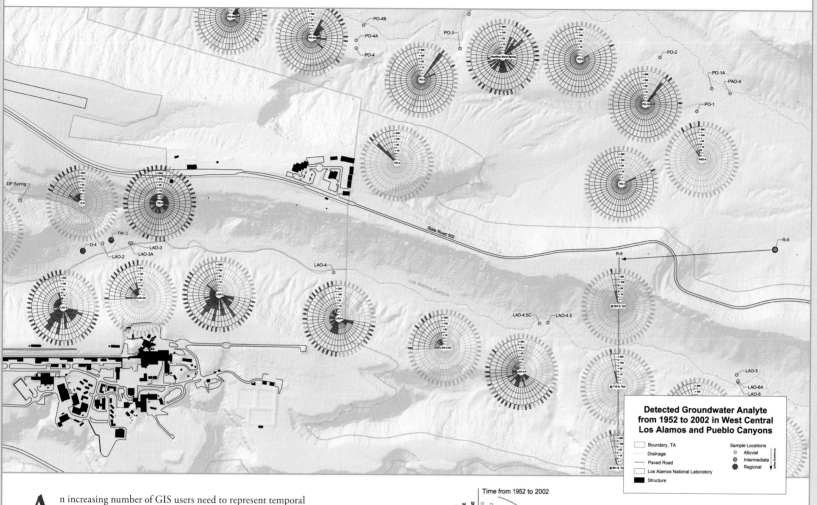

Detected Groundwater Analyte from 1952 to 2002 in West Central Los Alamos and Pueblo Canyons

Boundary, TA
Drainage
Paved Road
Los Alamos National Laboratory
Structure

Sample Locations
Alluvial
Intermediate
Regional

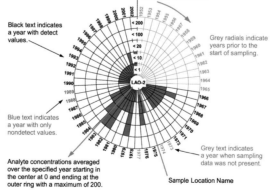

Time from 1952 to 2002

Black text indicates a year with detect values.

Grey radials indicate years prior to the start of sampling.

Blue text indicates a year with only nondetect values.

Grey text indicates a year when sampling data was not present.

Analyte concentrations averaged over the specified year starting in the center at 0 and ending at the outer ring with a maximum of 200.

Sample Location Name

LAO-2

An increasing number of GIS users need to represent temporal or time dependent data on maps in a meaningful and concise manner. The ability to analyze data for both spatial and temporal patterns in a single graphical presentation provides a powerful incentive to develop visualization tools for communication to map users and decision makers. Although there are many different methods to display scientific data to help discern either spatial or temporal trends, few visualization software packages allow for a single graphical presentation within a geographic context.

The authors' challenge was to display chemical concentration data for many different groundwater monitoring wells for a single chemical or analyte sample over time. Sample data had been collected for many years; some wells were sampled as early as 1950. The goal was to present this complex temporal data on a map so that one could easily visualize spatially the movement and change in concentration of the analyte, both horizontally and vertically, and to assess the potential for the analyte reaching the water table after many years.

The authors developed the clock diagram code and method for temporal and spatial visualization. The clock diagrams are analogous to "rose diagrams" often used to depict wind direction or strike direction in geologic applications. The clock diagram method consists of three steps: (1) sample location data, including depth, is compiled

along with attribute data such as concentration and time; (2) all data is read into a computer code, and clock diagram graphics are created; and (3) clock diagram graphics are placed on the map according to geographic location of the well where samples were collected.

Although this method of representing temporal data through the use of clock diagrams is not entirely new, the use of multiple clock diagrams as GIS symbology to emphasize movement of sampled data over time is a new concept. The clock diagram code and method have been proven to be an effective means in identifying temporal and spatial trends when analyzing groundwater data.

Los Alamos National Laboratory Earth and Environmental Science Division

Los Alamos, New Mexico, USA

By Chris Echohawk, Diana J. Hollis, Thomas L. Riggs, Doug E. Walther, and Marc S. Witkowski

Contact

Thomas Riggs

tlriggs@lanl.gov

Software

ArcMap 8.3, Microsoft Excel, and Windows 2000

Hardware

Compaq Professional Workstation SP750 Pentium III

Printer

HP Designjet 1055cm

Data Source(s)

Los Alamos National Laboratory

Natural Resources—Mining and Earth Science

Infiltration Studies

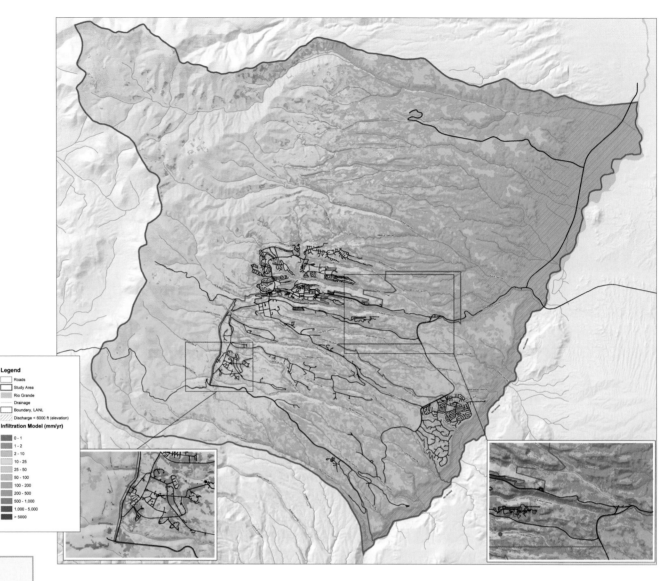

Legend
- Roads
- Study Area
- Rio Grande
- Drainage
- Boundary, LANL
- Discharge < 6000 ft (elevation)

Infiltration Model (mm/yr)
- 0 - 1
- 1 - 2
- 2 - 10
- 10 - 25
- 25 - 50
- 50 - 100
- 100 - 200
- 200 - 500
- 500 - 1,000
- 1,000 - 5,000
- > 5000

Los Alamos National Laboratory Earth and Environmental Science Division

Los Alamos, New Mexico, USA
By Kay H. Birdsell, James W. Carey,
Gregory L. Cole, Edward M. Kwicklis,
Roger P. Prueitt, Douglas E. Walther, and
Marc S. Witkowski

Contact
Marc Witkowski
witk@lanl.gov

Software
ArcGIS Spatial Analyst, ArcMap 8.3, and
Windows 2000

Printer
HP Designjet 1055cm

Data Source(s)
Los Alamos National Laboratory

Natural Resources—Mining and Earth Science

Infiltration Scenario for the Pajarito Plateau, Northern New Mexico

A map of net infiltration for the Pajarito Plateau, Northern New Mexico, was created for the pre-Cerro Grande fire period using new and previously published estimates of point infiltration in upland areas as well as estimates of stream flow losses and gains along canyon bottoms. The point infiltration estimates are based on a combination of techniques that include the use of the Richards' equation, the chloride mass balance method, and numerical modeling.

The infiltration rates estimated with these techniques were extrapolated to uncharacterized parts of the study area using maps of environmental variables, which were correlated with infiltration and spatial algorithms implemented with GIS software using the mapped variables. This map indicates that infiltration rates on mesas of the Pajarito Plateau are generally less than two millimeters per year (mm/yr), except near faults. Infiltration rates at higher elevations in the Sierra de los Valle are typically greater than 25 mm/yr in mixed conifer areas and greater than 200 mm/yr in areas vegetated by aspen.

An irregular transition zone with infiltration rates between 10 and 25 mm/yr exists near the western edge of the Pajarito Plateau adjacent to the Sierra de los Valle. Canyon bottom infiltration rates are highly variable, ranging from several hundred mm/yr to several mm/yr. The total net infiltration of approximately 8,600 acre-feet per year is consistent with estimates of the steady state of groundwater discharge to perennial streams in the study area.

Limitations of the study are that it does not address the effects of the Cerro Grande fire on the hydrology of the study area, nor does it completely capture the complex and sometimes incompletely documented history of laboratory-generated discharges during its 60-year history.

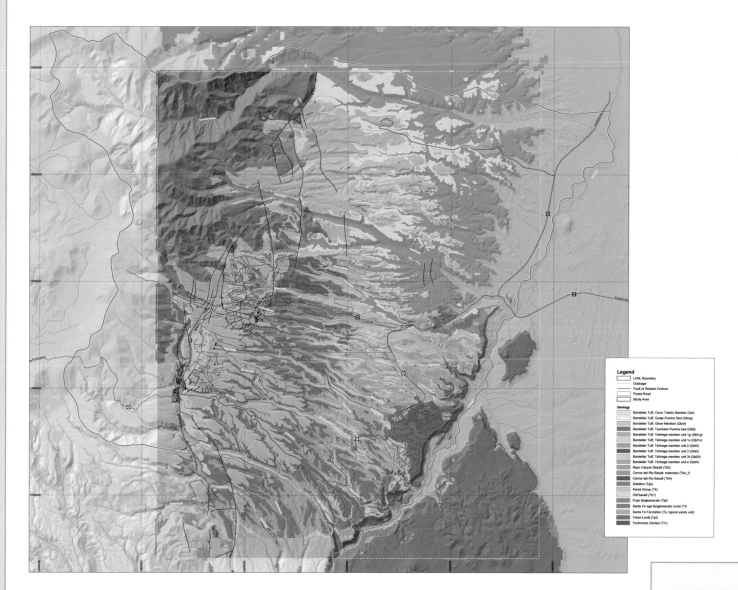

Legend
LANL Boundary
Drainage
Fault or Related Feature
Paved Road
Study Area

Geology
Bandelier Tuff, Cerro Toledo Member (Qct)
Bandelier Tuff, Guaje Pumice Bed (Qbog)
Bandelier Tuff, Otowi Member (Qbof)
Bandelier Tuff, Tsankawi Pumice bed (Qbt)
Bandelier Tuff, Tshirege member unit 1g (Qbt1g)
Bandelier Tuff, Tshirege member unit 1v (Qbt1v)
Bandelier Tuff, Tshirege member unit 2 (Qbt2)
Bandelier Tuff, Tshirege member unit 3 (Qbt3)
Bandelier Tuff, Tshirege member unit 3t (Qbt3t)
Bandelier Tuff, Tshirege member unit 4 (Qbt4)
Bayo Canyon Basalt (Tb2)
Cerros del Rio Basalt, extended (Tb4_f)
Cerros del Rio Basalt (Tb4)
Galisteo (Tgs)
Keres Group (Tk)
Old basalt (Tb1)
Puye fanglomerate (Tpf)
Santa Fe age fanglomeratic rocks (Tf)
Santa Fe Formation (Ts, typical sandy unit)
Totavi Lentil (Tpt)
Tschicoma Dacites (Ttt)

Surface Geology for the Los Alamos Regional Infiltration Study Area

Geology shown for the Los Alamos infiltration study area was created by carefully matching and then combining regional scale as well as local site scale geologic models of different resolutions. The dashed box represents the boundary between the two models. The surface geology shown is used as one layer of input into a study of surface water infiltration rates in the region surrounding the Los Alamos National Laboratory. The geology shown is a combination of a coarse resolution regional geologic model and a finer resolution model more focused on the Los Alamos National Laboratory.

Both models received the same legend classification with symbols intended to create an appearance that is easy to look at while maximizing the contrast between adjacent units to aid in identifying boundaries. Geologic interpretation is further enhanced by the overlay of a semitransparent hillshade along with drainage channels and fault lines. Canyon names, transportation infrastructure, and the study unit boundary are added to assist in defining the locations of units of interest.

Los Alamos National Laboratory Earth and Environmental Science Division
Los Alamos, New Mexico, USA
By Kay H. Birdsell, James W. Carey, Gregory L. Cole, Edward M. Kwicklis, Roger P. Prueitt, Douglas E. Walther, and Marc S. Witkowski

Contact
Douglas E. Walther
walther@lanl.gov

Software
ArcGIS Spatial Analyst, ArcMap 8.3, and Windows 2000

Printer
HP Designjet 1055cm

Data Source(s)
Los Alamos National Laboratory

Natural Resources—Mining and Earth Science

100+ Years of Land Change for Coastal Louisiana

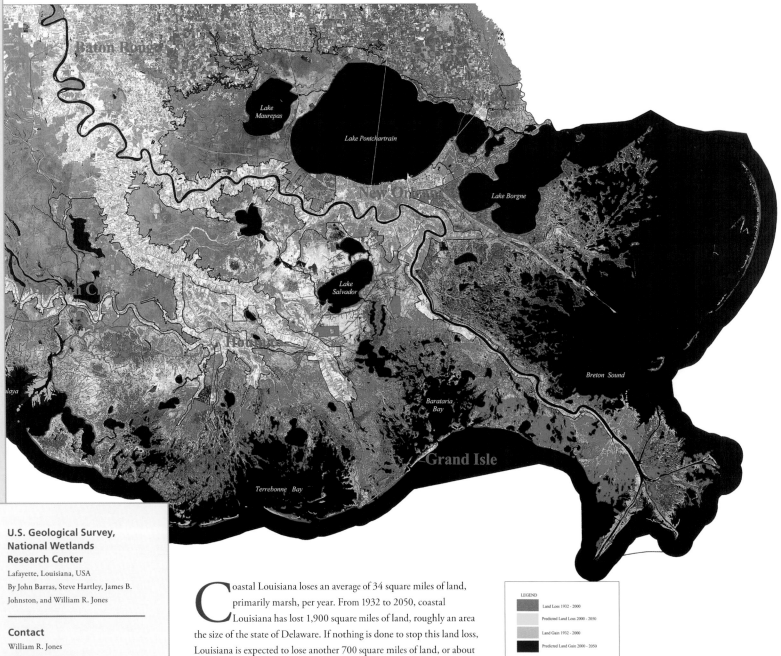

U.S. Geological Survey, National Wetlands Research Center

Lafayette, Louisiana, USA

By John Barras, Steve Hartley, James B. Johnston, and William R. Jones

Contact

William R. Jones

william_jones@usgs.gov

Software

ArcView Image Analysis and Windows NT

Printer

HP Designjet 3500cp

Data Source(s)

Britsch, L.D., and Dunbar, J.B., 1993, "Land-loss Rates Louisiana Coastal Plain," *Journal of Coastal Research* v.9, p. 324–338; Cowardin et al, 1979, "Classification of Wetlands and Deepwater Habitats of the United States" -79/31; and 2000 Thematic Mapper data

Natural Resources—Mining and Earth Science

Coastal Louisiana loses an average of 34 square miles of land, primarily marsh, per year. From 1932 to 2050, coastal Louisiana has lost 1,900 square miles of land, roughly an area the size of the state of Delaware. If nothing is done to stop this land loss, Louisiana is expected to lose another 700 square miles of land, or about equal to the size of the greater Washington, D.C.–Baltimore area, during the next 50 years. Louisiana accounted for an estimated 90 percent of the coastal marsh loss in the lower 48 states during the 1990s.

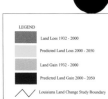

LEGEND

Land Loss 1932 - 2000

Predicted Land Loss 2000 - 2050

Land Gain 1932 - 2000

Predicted Land Gain 2000 - 2050

Louisiana Land Change Study Boundary

Map of Surficial Deposits and Materials in the Eastern and Central United States (East of 102° West Longitude)

This map depicts the aerial distribution of surficial geologic deposits and other materials that accumulated or formed during the last two or more million years. That period includes all activities of the human species. These materials are at the surface of the Earth and make up the ground on which we walk, the dirt in which we dig foundations, and the soil in which we grow crops. The map is based on 31 published maps in the U.S. Geological Survey's Quaternary Geologic Atlas of the United States (U.S. Geological Survey Miscellaneous Investigations Series I–1420). It was compiled at 1:1,000,000 scale, to be viewed as a digital map at a nominal scale of 1:2,000,000, and to be printed as a conventional paper map at 1:2,500,000 scale.

Each 1:1,000,000 4° x 6° quadrangle map in the atlas was simplified. Because the quadrangles of the atlas were compiled and printed in different projections and on different bases, the projections and bases were converted to a common one for publication. First, the source geology had to be recompiled to match the digital base on which the map was to be printed. The GIS file of streams and water bodies from the National Atlas of the United States was clipped into 4° x 6° quadrangles to match the original maps and converted to the projections of the original maps. The geology was then compiled on these drainage bases.

The new, hand-drafted compilations of the geology were scanned, vectorized with the LT4X computer program, and converted into ArcInfo coverages. Polygons and lines were attributed, and map data for the individual quadrangles was unprojected to geographic coordinates, appended to one another, and edgematched. Selected shorelines, lakes, and rivers were added from the hydrographic coverage. The geology was reconciled along the borders of the adjacent quadrangles, and the entire map was converted to the Lambert azimuthal equal area projection.

The map layout was produced by importing ArcInfo shapefiles for the geologic database and planimetric base into Adobe Illustrator through the plug-in Avenza MAPublisher. The map sheet is accompanied by a 46-page pamphlet of detailed map unit descriptions. The database is available as ArcInfo export files and ArcView shapefiles at http://pubs.usgs.gov/imap/i-2789/.

U.S. Geological Survey
Denver, Colorado, USA
By David S. Fullerton

Digital database by Charles A. Bush and Jean N. Pennell

Edited by Diane E. Lane

Digital cartography by Diane E. Lane, Nancy Shock, and William Sowers

---

Contact
Diane E. Lane
delane@usgs.gov

Software
ArcInfo, ArcView, Adobe Illustrator 8.0, Avenza MAPublisher, and Infotec LT4X

Hardware
Intel PC, Macintosh, and UNIX

Printer
HP Designjet 3500 and four-color offset printing

Data Source(s)
Quaternary Geologic Atlas of the United States

Natural Resources—Mining and Earth Science

Seismicity Maps of the Santa Rosa Quadrangle, California, for 1969–1995

Significant Earthquake Clusters

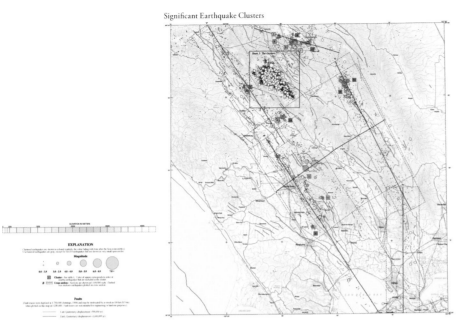

Seismicity

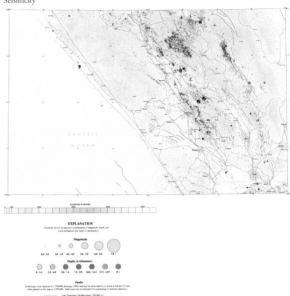

This map set depicts several aspects of recorded seismicity that occurred from 1967 through 1995 within the Santa Rosa, California, quadrangle. The seismicity map and the related cross sections illustrate the locations, depths, and magnitudes of earthquakes. Representative focal mechanisms are also plotted to indicate the fault orientation and direction of motion on many faults. The earthquake clusters map identifies the most significant temporal clusters of earthquakes illustrating the time dependent properties of earthquake occurrence.

The Geysers geothermal area map shows the time dependent relation between the inception of geothermal power production and the occurrence of earthquake activity. In concentrations of earthquakes located beneath mapped faults, the map authors infer that the seismicity defines the subsurface orientation of these faults. Even though some concentrations of earthquakes do not underlie mapped faults, they also infer that these concentrations represent active but unnamed faults. Earthquakes also occur as isolated events beneath mapped faults and throughout the region.

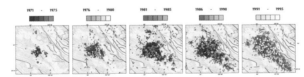

Time Dependent Seismicity in the Geysers Area

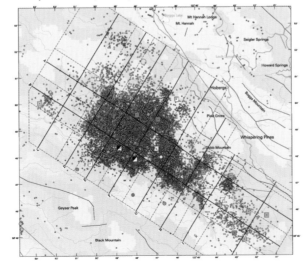

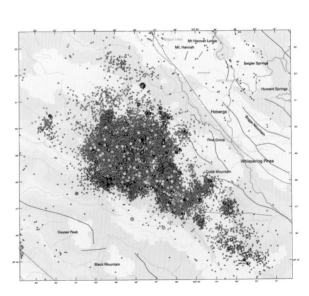

U.S. Geological Survey

Menlo Park, California, USA

By David H. Oppenheimer,
Jerome S. Preiss, and Stephen R. Walter

Contact

Stephen R. Walter

swalter@usgs.gov

Software

ArcInfo 7.2

Hardware

Sun Ultra A60

Printer

HP Designjet 2500cp

Data Source(s)

Northern California Seismic Network

Natural Resources—Mining and Earth Science

Sub-Andean Geological Map

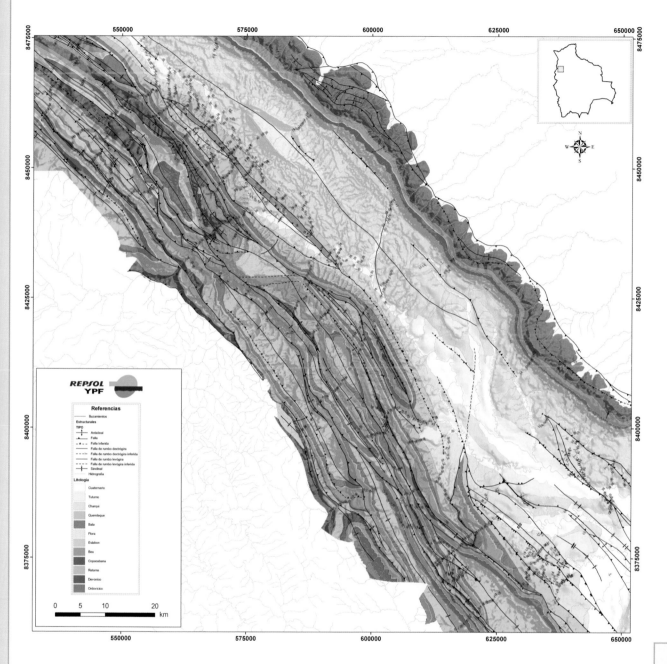

A 10-meter resolution digital elevation model was obtained from SPOT stereo imagery. SPOT panchromatic (10-meter resolution) and Landsat 7 Enhanced Thematic Mapper (15-meter resolution) imagery was used for interpretation and mapping of lithological contacts and structures. Topology was built on contacts to obtain the formational units (polygons) with ArcInfo, and the corresponding attributes were loaded. Dip direction data obtained in the field was also added.

With data entry completed, symbology was assigned to each layer. The resulting surface geology map was used as a source of information for structural analysis combined with the topographic information.

Repsol YPF

Buenos Aires, Argentina
By Gabriel Alvarez and Bernardo Troncaro

Contact

Omar Baleani
obaleani@aeroterra.com

Software

ArcInfo 8.3

Printer

HP 2500c

Data Source(s)

Data dips, digital elevation model, and satellite imagery

Natural Resources—Mining and Earth Science

The Application of GIS to Bauxite Mining in Jamaica

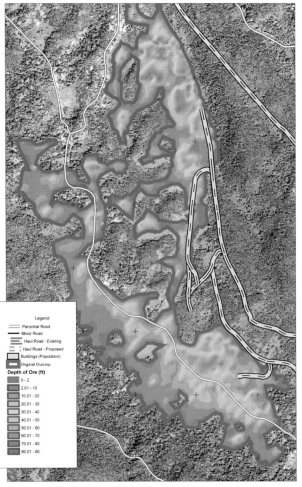

Map 1: Bauxite Thickness Isopach Deposit 135—Original Model

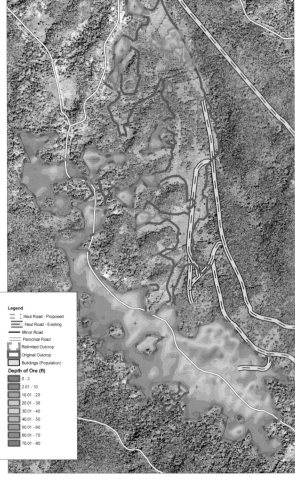

Map 2: Bauxite Thickness Isopach Deposit 135—Recoverable Model

Alpart Mining Venture

Manchester, Jamaica
By Deborah Almarales–Hamm,
Gil de Campos, Bryan Murray, and
Natalie Russell

Contact

Gil de Campos
gil.decampos@kaiseral.com

Natalie Russell
natalie.russell@kaiseral.com

Software

ArcGIS 8.3, ArcGIS Spatial Analyst,
ArcInfo, ArcSDE, ArcView, ArcGIS
Geostatistical Analyst, Vulcan 3D, and
Windows 2000

Hardware

Intel Xeon processor

Printer

HP Designjet 3500cp

Data Source(s)

Field survey, drilling, sample analyses,
and orthorectified aerial photography and
satellite imagery

**Natural Resources—Mining
and Earth Science**

Jamaica is the world's fourth largest producer of bauxite, the primary mineral used in the production of alumina, with bauxite and alumina production ranking as the country's second largest foreign currency earner.

The island is partially covered by tertiary limestone, containing bauxite deposits in irregular and scattered paleokarstic channels relatively close to the surface. Although standard surface mining techniques are used, mining operations are constrained by rugged terrain, population density, varying ore quality, and small deposit sizes, coupled with outdated and inflexible data and maps. This operationally challenging environment prompted the need for development and implementation of cost-effective approaches to bauxite reserves utilization and management. GIS applications supported by satellite imagery and orthorectified aerial photography were implemented to manage, analyze, and display data on tonnage, ore quality, location, and ownership.

The Application of GIS to Bauxite Mining in Jamaica

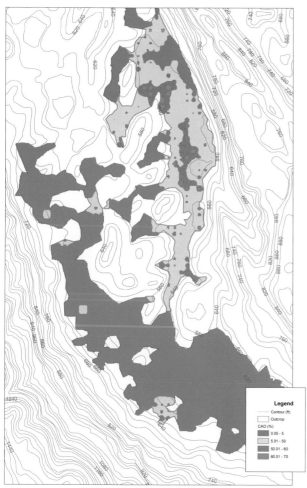

Map 3: Distribution of CaO Within Deposit 135—Original Model

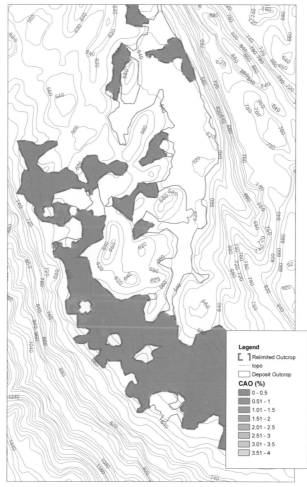

Map 4: Deposit 135 Detailing Mineable Areas Based on the Distribution of CaO

GIS applications, using ArcGIS 8.3 and its extensions and Vulcan, a customized 3D modeling and mine planning software, provide tools that assist in the decision making process and reserves management. This results in improved ore recovery control by enabling manipulation of various mine planning scenarios and constraints in the development and execution of mine plans. Maps 1 and 3 illustrate models of the thickness isopach and calcium oxide (CaO) distribution for a deposit using all collar (drill depth) and assay (quality) data. By applying criteria such as quality constraints on the deposit, the recoverable tonnage and quality are estimated and ore recovery strategy developed. Maps 2 and 4 illustrate the thickness isopach and CaO distribution respectively and where CaO was constrained for the actual recoverable areas within the deposit.

Alpart Mining Venture

Manchester, Jamaica

By Deborah Almarales–Hamm, Gil de Campos, Bryan Murray, and Natalie Russell

Contact

Gil de Campos
gil.decampos@kaiseral.com

Natalie Russell
natalie.russell@kaiseral.com

Software

ArcGIS 8.3, ArcGIS Spatial Analyst, ArcInfo, ArcSDE, ArcView, ArcGIS Geostatistical Analyst, Vulcan 3D, and Windows 2000

Hardware

Intel Xeon processor

Printer

HP Designjet 3500cp

Data Source(s)

Field survey, drilling, sample analyses, and orthorectified aerial photography and satellite imagery

Natural Resources—Mining and Earth Science

Czech Geological Survey—
Using the National Geodatabase GEOCR50

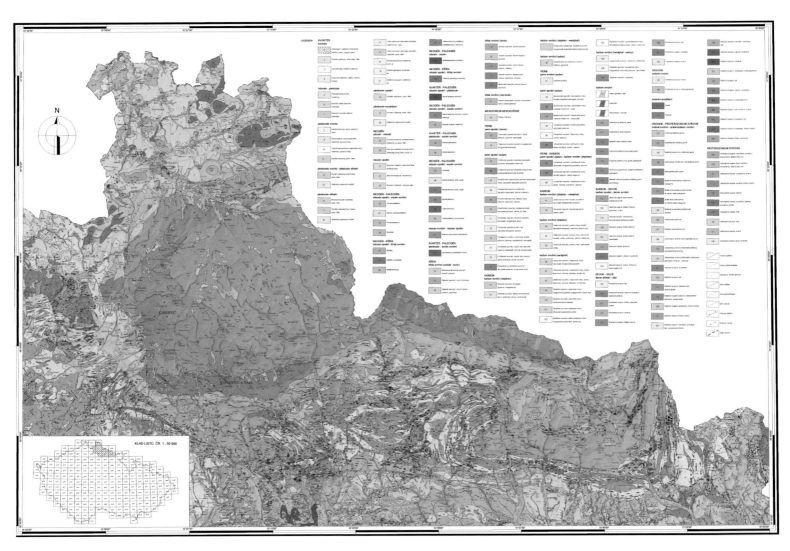

Czech Geological Survey

Prague, Czech Republic

By Vladimir Ambrozek, Veronika Kopackova, Zuzana Krejci, and Robert Tomas

Contact

Veronika Kopackova

kopackova@cgu.cz

Software

ArcGIS 3D Analyst 8.3, ArcSDE 8.3, and Windows 2000

Printer

HP Designjet 1055cm

Data Source(s)

Czech Geological Survey

Natural Resources—Mining and Earth Science

In 1994, the Czech Geological Survey (CGS) began to use GIS technology extensively to meet the increasing demand for digital information about the environment. GIS in the CGS is focused on the methods of spatial data processing, unification, and dissemination. Digital processing of geological and applied maps and the development of GIS follows standardized procedures using common geological dictionaries and graphic elements. Recently, the main objective has been to create and implement a uniform geological data model (geodatabase) and provide the public and the scientific community with easy access to geodata, such as CGS' Map Server (www.geology.cz), via the World Wide Web.

The process of digitizing geological maps at 1:50,000 scale was part of a CGS project. The conversion of geological maps into digital form was divided into several steps—scanning of the source material for vectorization, transformation of the image projection system, and vectorization. Digitizing 214 map sheets of geological maps covering the entire Czech Republic was finished at the end of 1997. The implementation of the vector maps into the GEOCR50 database was completed in 1998.

GEOCR50 is a unique geographic information system, containing more than 260,000 mapped geological objects from the entire Czech Republic. The fundamental part of this geodatabase is the unified national geological index (legend), which consists of four main types of information—chronostratigraphical units, regional units, lithostratigraphical units, and lithological description of rocks. The geodatabase has been under revision since 1998 leading to the creation of a seamless digital geological map of the Czech Republic. This geodatabase has already been used for land use planning by government and local authorities.

The geological map of the Krkonose–Jizera mountains is a cartographic presentation of one part of the GEOCR50 geodatabase.

Brazil Petroleum Infrastructure

Offshore Drilling Platform

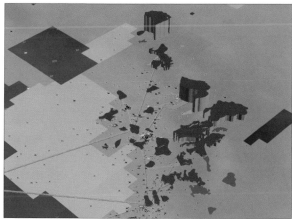

Detailed Offshore Producing Field

Regional Overview of Southern Brazil and Continental Shelf

Global View of South America

ArcGlobe™ has the capability to put many data sets in perspective. This satellite view of the Rio de Janeiro–São Paulo region of southern Brazil appears as a silk screen—the world topography and culture are the raster backdrop. Concessions, platforms, and pipelines are drawn on top as vectors. They show the offshore location of exploration and production areas and the onshore gathering points of refinery and distribution along the coast.

The halftone geology delineates the onshore Parana and São Francisco basins and the offshore Santos, Campos, and Espirito Santo oil producing basins that lie along the clearly visible continental shelf. Insets add a global view with cloud cover, a close-up view of an offshore oil field, and a perspective view of a drilling platform under clouds. Such synoptic views help convey complex and interrelated information to the public, investors, agencies at large, and in-house colleagues and management.

ESRI
Meteorlogix, LLC
Tobin International
Redlands, California, USA
By Paul Hackleman, Tobin International;
Clive Reese, Meteorlogix; and Andrew
Zolnai, ESRI

Contact
Andrew Zolnai
azolnai@esri.com

Software
ArcGIS 3D Analyst, ArcGIS 3D Analyst,
and ArcInfo

Hardware
Dell Latitude C840

Data Source(s)
ESRI, IHS Energy, Meteorlogix, and Tobin
International

Natural Resources—Petroleum

Prospect Hydroelectric Project, Rogue River, Oregon

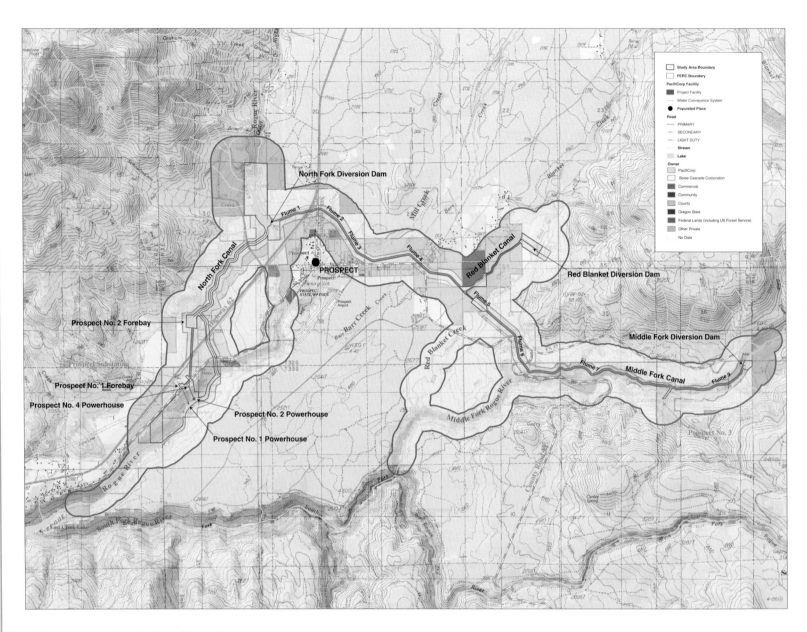

ENTRIX, Inc.

Seattle, Washington, USA
By Zoltan Der and Shruti Mukhtyar

Contact

Shruti Mukhtyar
smukhtyar@entrix.com

Software

ArcInfo 8.3 and Windows 2000

Printer

HP Designjet 1055cm

Data Source(s)

ESRI; ENTRIX, Inc.; Jackson County,
Oregon; State of Oregon; PacifiCorp; and
U.S. Geological Survey

Natural Resources—Water

The project is located in Jackson County, Oregon, on the Rogue River and on two tributaries approximately 45 miles northeast of Medford, Oregon, near the town of Prospect. Lands abutting the project include lands owned by the U.S. Forest Service, state of Oregon, Boise Cascade, PacifiCorp, and other private landowners.

Construction took place between 1911 and 1946 by the California–Oregon Power Company (Copco). Copco subsequently merged with Pacific Power and Light Company in 1961. Since then, the project has been owned and managed by PacifiCorp. It includes three concrete diversion dams, three powerhouses, and a water conveyance system of approximately 9.3 miles. The entire hydroelectric system is a run-of-the-river operation, and PacifiCorp is applying to the Federal Energy Regulatory Commission (FERC) for a new 50-year license to continue operation.

The project collectively diverts up to 1,050 cubic feet per second of water from the three diversion dams. Water is transported through the water conveyance system of canals and flumes to a small (15-acre-foot) forebay. Water then is routed through powerhouses and is returned to the Rogue River approximately two miles upstream of Lost Creek Lake.

ENTRIX, Inc., has worked closely with the computer architecture group at PacifiCorp to integrate data sources from the client, which include property ownership and detailed project facilities, with data from other state and federal agencies. The map products are an integral component of the draft license application and have been designed to FERC specifications. ENTRIX staff scientists and PacifiCorp personnel have extensively used the maps for field surveys and studies. All spatial data processed or generated for the project by ENTRIX is being transferred in geodatabase format with full metadata to PacifiCorp for integration into its enterprise GIS database.

California Water—21st Century Gold

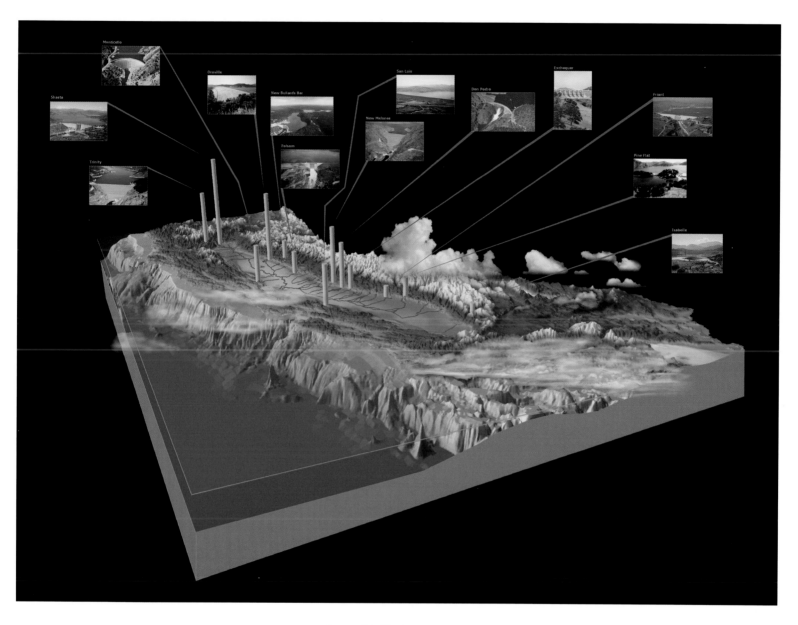

Dams of California's Central Valley Symbolized by Storage Capacity

This poster includes dams in California's Central Valley with maximum storage capacities of greater than 500,000 acre-feet. Dam location symbols have been extruded by maximum storage capacity. Total storage behind these facilities exceeds 25.5 million acre-feet. One acre-foot equals 326,000 gallons—enough water to supply a five-person household for an entire year.

U.S. Bureau of Reclamation
Sacramento, California, USA
By Tom Heinzer and Diane Williams

Contact
Tom Heinzer
theinzer@mp.usbr.gov

Software
ArcMap, ArcGIS 3D Analyst, Photoshop, and Windows XP

Printer
HP Designjet 5500

Data Source(s)
U.S. Bureau of Reclamation and U.S. Geological Survey

Natural Resources—Water

Township of Michipicoten Municipal Groundwater Study—Quaternary Geology

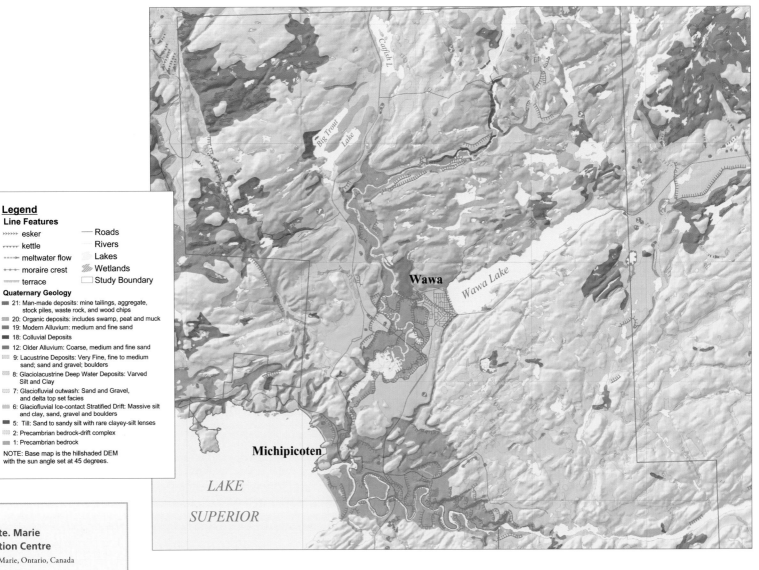

Legend

Line Features

- ↠↠↠ esker
- ⌄⌄⌄⌄ kettle
- ·····» meltwater flow
- ◂──▸ moraire crest
- ⎍⎍⎍ terrace

- —— Roads
- —— Rivers
- ▢ Lakes
- ⫽⫽ Wetlands
- ▢ Study Boundary

Quaternary Geology

- 21: Man-made deposits: mine tailings, aggregate, stock piles, waste rock, and wood chips
- 20: Organic deposits: includes swamp, peat and muck
- 19: Modern Alluvium: medium and fine sand
- 18: Colluvial Deposits
- 12: Older Alluvium: Coarse, medium and fine sand
- 9: Lacustrine Deposits: Very Fine, fine to medium sand; sand and gravel; boulders
- 8: Glaciolacustrine Deep Water Deposits: Varved Silt and Clay
- 7: Glaciofluvial outwash: Sand and Gravel, and delta top set facies
- 6: Glaciofluvial Ice-contact Stratified Drift: Massive silt and clay, sand, gravel and boulders
- 5: Till: Sand to sandy silt with rare clayey-silt lenses
- 2: Precambrian bedrock-drift complex
- 1: Precambrian bedrock

NOTE: Base map is the hillshaded DEM with the sun angle set at 45 degrees.

Sault Ste. Marie Innovation Centre

Sault Ste. Marie, Ontario, Canada

By Marlene McKinnon

Contact

Marlene McKinnon

mmckinnon@ssmic.com

Software

ArcInfo 8.1, ArcView 3.2, and Windows 2000 Professional

Hardware

Dell Pentium 4

Printer

HP Designjet 1055cm

Data Source(s)

Corporation of the Township of Michipicoten, Ontario Ministry of Environment, Ontario Ministry of Natural Resources, and Ontario Ministry of Northern Development and Mines

Natural Resources—Water

The Township of Michipicoten is one of 31 communities participating in the Ontario Ministry of Environment's Municipal Groundwater Studies. These studies were initiated in August 2001, following the Walkerton, Ontario, water contamination in May 2000.

Approximately three million Ontario residents, comprising more than 200 communities, rely on municipal groundwater-based systems to provide safe drinking water to service the residential, industrial, commercial, and institutional sectors. Of the rural population, approximately 90 percent rely on groundwater for drinking and other uses. The provincial Clean Water initiative characterizes hydrogeological conditions, of which quaternary geology is one component. Quaternary geology consists of rock formations from the beginning of the last ice age two million years ago to the present. The initiative will also focus on the development of immediate and long-term groundwater protection plans for municipalities across Ontario using groundwater as their water supply source.

The quaternary geology polygon layer for this map was obtained from the Ontario Ministry of Northern Development and Mines CAD format files. Where data was not available at 1:50,000 scale, a clip of the provincial quaternary geology at a 1:1 million scale was incorporated into the extracted shapefile (southern portion of the study area). David Sawicki, professional engineer and senior geological engineer of Morrison Environmental Limited, performed the verification of quaternary geological unit identification and legend description.

The quaternary geology information for the Township of Michipicoten Municipal Groundwater Studies Report, with other municipal groundwater study reports, will generate a provincial image to assist in the development of future environmental policies and effective groundwater protection strategies.

Aeronautical Charting in New Zealand

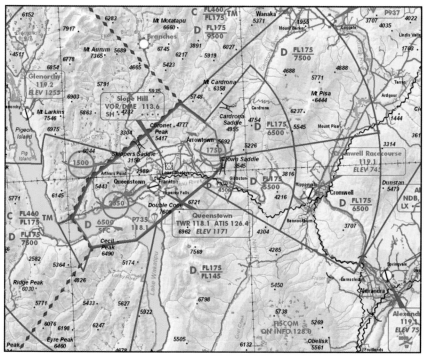

Charting 1:500,000

Explorer Graphics Ltd. (EGL) produced a new national coverage of large-format aeronautical Visual Navigation and Planning Charts (20) for its clients, Airways Corporation and Civil Aviation Authority of New Zealand.

Specification modifications and database enhancements resulted in the creation of a new generation of charts produced entirely with ArcMap. Production of these charts is based on the requirement to manage and update aeronautical data in a timely and effective manner from a central database. All data, including context topographical information, is managed within a personal geodatabase. Automated update procedures and special use of ArcGIS feature-linked annotation are essential to maintain data integrity and the accuracy of information representation.

Custom symbol sets have been designed, and Visual Basic development has automated map information presentation, the management of multiple data frames, and map templates for the creation of export files for offset print production. EGL plans to extend the current map production methodologies for the generation of aeronautical en route charts using the same methodology. These charts will use the same relevant aeronautical data for consistency, standardization of specifications, and database integrity.

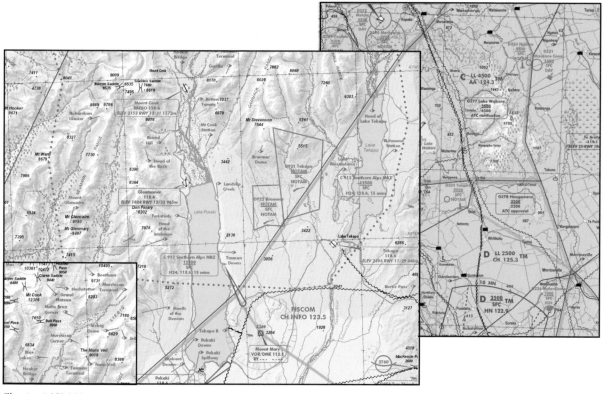

Charting 1:250,000

Explorer Graphics Ltd.

Porirua, New Zealand

By Gael Cutress, Beryl Pimblott, Carol Seymour, and Yongji Zhang

Contact

Beryl Pimblott

beryl@egl.co.nz

Software

ArcGIS 8.2 and Windows 2000

Printer

Offset printing

Data Source(s)

Airways Corporation, Civil Aviation Authority, GeographX, and Land Information New Zealand

Natural Resources—Water

Three-Dimensional Visualization of Cod Spatial Dynamics and Vertical Movements in European Waters

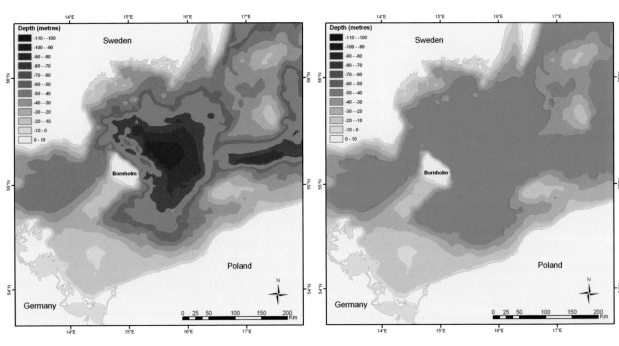

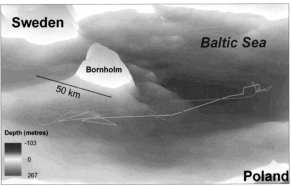

Danish Institute for Fisheries Research

Charlottenlund, Denmark

By Kerstin Geitner and Stefan Neuenfeldt

Contact

Kerstin Geitner

kjg@dfu.min.dk

Software

ArcView 8.3, ArcGIS 3D Analyst, ArcGIS Tracking Analyst, and Macromedia Flash MX Professional 2004

Hardware

Dell Latitude D800

Printer

Aficio Color 4006

Data Source(s)

Baltic Sea Research Institute Warnemünde and electronic data storage tags

Natural Resources—Water

Cod stocks in European waters are in serious decline recently, drawing media coverage to that subject. Danish Institute for Fisheries Research (DIFRES) participated in a television program about threatened and endangered fish species on the national Danish Channel DR2 shown in March 2004. The program focuses on biologists' work in estimating stock size for different fish species. One of the items produced for the show by DIFRES is a video clip in which ArcGIS 3D Analyst was used to show the movements of a single tagged cod.

The three-dimensional picture shows the movement of a single cod for 55 days from April to June 2003. The cod was tagged with an electronic data storage tag (green dot on picture) and caught (red dot on picture) at locations approximately 27 kilometers from each other. The three-dimensional visualization illustrates the movement of the cod showing that the cod mainly stayed in two different locations: one location to the southeast of Bornholm where it was found in the water column and another location to the southwest of Bornholm where it was found at the bottom. The scientific interpretation of these results is that the cod was probably spawning in the water column while residing in the first area and feeding at the bottom while residing in the second area.

The full video clip can be seen on the DIFRES home page at www.difres.dk. There is also another video clip produced with ArcGIS Tracking Analyst for the same television program. This clip shows the distribution area of cod in the Bornholm Basin in the Baltic Sea since the 1950s.

Two small maps explicitly show the geographic extent of the bottom area covered by water with an oxygen concentration greater than two milligrams per liter and a salinity greater than 11 in the Bornholm Basin of the Baltic Sea during the 44 years that the video clip covers. Biological surveys along with measurements of physical and chemical parameters indicate that these conditions, highlighted by the bright green area on the maps, characterize the main cod distribution area in this region. In 1971, the area for cod distribution was at its lowest with an area less than 9,000 square kilometers. In 1968 and 1974 the area was at its largest extension with approximately 33,000 square kilometers. The average for the 44 measured years is approximately 20,000 square kilometers per year.

All three maps are in the Mercator projection, and the three-dimensional map was produced with a vertical exaggeration of 200X.

Tauranga Harbour Tidal Movements

Legend
(Not to Scale)

High Tide—1.5 Hours

This is a numerical model of Tauranga Harbour during spring tidal conditions, which shows a snapshot in time of a dynamic data set. The estuary model has been shown to predict the currents and sea levels accurately when compared to highly detailed measurements. With this information, a resource planner can view and use current and sea level information to respond to applications for coastal development, fisheries, and other environmental assessments. Approximately 1,300 vessels call at the Port of Tauranga annually, moving 12,104,000 tons of cargo. Should an oil spill occur, this information will assist environmental and marine scientists in assessing the flushing movements of the tides.

These maps were created using ArcMap software's multiple attributes symbology, and each point displays information for speed, depth, and direction using color to show depth, size to show speed, and azimuth to show direction.

This project was a winner in the international Hewlett–Packard Best GIS Output competition held last year.

Low Tide—7.5 Hours

Environment Bay of Plenty
Whakatane, Bay of Plenty, New Zealand
By Ross Chrystall

Contact
Ross Chrystall
ross@envbop.govt.nz

Software
ArcMap

Printer
HP Designjet 5500

Natural Resources—Water

Water Quality Monitoring of the Italian Rivers, a GIS Approach

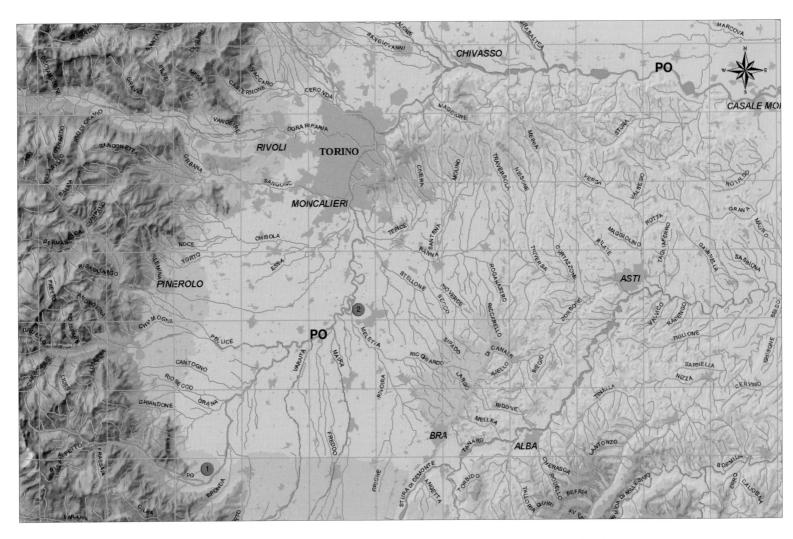

Environmental Protection National Agency

Rome, Italy

By Ernesto Consiglio and Roberto Durastante, ESRI Italia; Stefano Ursino, APAT

Contact

Stefano Ursino

ursino@apat.it

Software

ArcInfo 8.3, Microsoft Visio, and Windows 2000

Hardware

Acer PC Pentium 4

Printer

HP Designjet

Data Source(s)

National Institute of Statistics and orthophotography

Natural Resources—Water

This poster shows 2002 water quality monitoring data of the Po River in the Piemonte region near Turin City. It is linked to the Italian river's shapefile with dynamic segmentation.

The route system of the Po River was created along with the points representing the monitoring stations. The linear events that represent the monitored features of the river were included and note whether the data is in compliance with the law (yes/no). The data about the water quality was ranked from 1 (not polluted, very good) to 5 (highly polluted, very bad).

Tables are linked to the monitored features with the measurements of the polluting elements such as pH, chlorine, lead, and temperature. Italian regions send these measurements from the monitoring stations to the Environmental Protection National Agency.

1. Monitored Features of the Po River: Sanfront

2. Monitored Features of the Po River: Carmagnola

The Watershed Fragmentation by Dams and Its Impacts on Freshwater Fishes

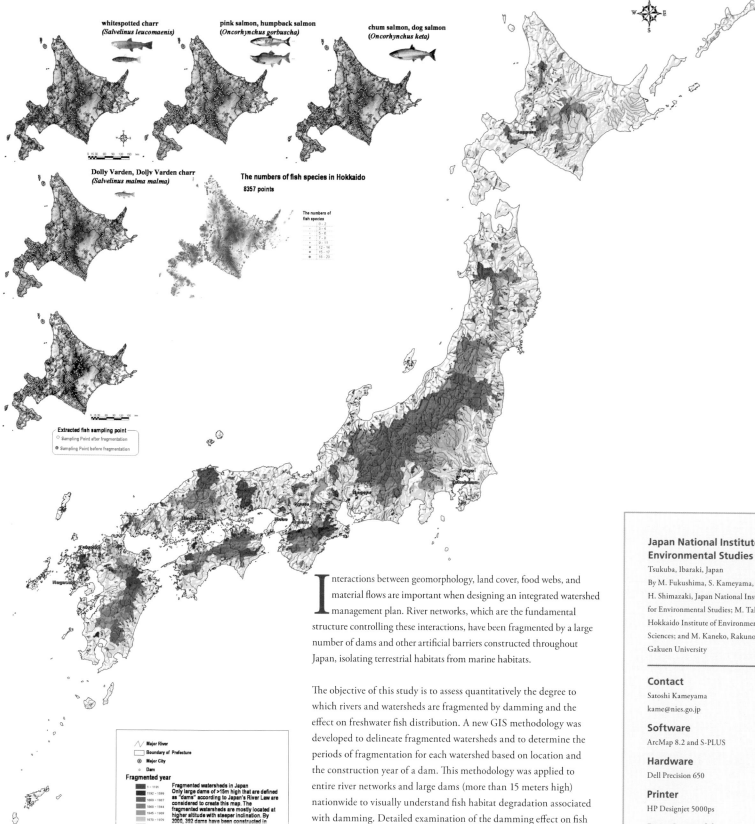

whitespotted charr
(*Salvelinus leucomaenis*)

pink salmon, humpback salmon
(*Oncorhynchus gorbuscha*)

chum salmon, dog salmon
(*Oncorhynchus keta*)

Dolly Varden, Dolly Varden charr
(*Salvelinus malma malma*)

The numbers of fish species in Hokkaido

8357 points

The numbers of
fish species
- 0 - 2
- 3 - 4
- 5 - 6
- 7 - 8
- 9 - 11
- 12 - 14
- 15 - 17
- 18 - 23

Extracted fish sampling point
- ○ Sampling Point after fragmentation
- ● Sampling Point before fragmentation

Major River
Boundary of Prefecture
⊙ Major City
○ Dam

Fragmented year
- 1 - 1191
- 1192 - 1599
- 1600 - 1867
- 1868 - 1944
- 1945 - 1969
- 1970 - 1979
- 1980 - 1989
- 1990 - 2000
- 2001 - 2010

Fragmented watersheds in Japan
Only large dams of >15m high that are defined
as "dams" according to Japan's River Law are
considered to create this map. The
fragmented watersheds are mostly located at
higher altitude with steeper inclination. By
2000, 392 dams have been constructed in
Japan. Two hundred and ninety-five more
dams are scheduled to be constructed in the
future.

Interactions between geomorphology, land cover, food webs, and material flows are important when designing an integrated watershed management plan. River networks, which are the fundamental structure controlling these interactions, have been fragmented by a large number of dams and other artificial barriers constructed throughout Japan, isolating terrestrial habitats from marine habitats.

The objective of this study is to assess quantitatively the degree to which rivers and watersheds are fragmented by damming and the effect on freshwater fish distribution. A new GIS methodology was developed to delineate fragmented watersheds and to determine the periods of fragmentation for each watershed based on location and the construction year of a dam. This methodology was applied to entire river networks and large dams (more than 15 meters high) nationwide to visually understand fish habitat degradation associated with damming. Detailed examination of the damming effect on fish distribution was conducted exclusively for the rivers in Hokkaido where a database of existing fish sampling data is available.

Japan National Institute for Environmental Studies
Tsukuba, Ibaraki, Japan
By M. Fukushima, S. Kameyama, and H. Shimazaki, Japan National Institute for Environmental Studies; M. Takada, Hokkaido Institute of Environmental Sciences; and M. Kaneko, Rakuno Gakuen University

Contact
Satoshi Kameyama
kame@nies.go.jp

Software
ArcMap 8.2 and S-PLUS

Hardware
Dell Precision 650

Printer
HP Designjet 5000ps

Data Source(s)
National GIS Data of Japan and original data

Natural Resources—Water

Using GIS for Data Integration and Visualization of a Deepwater Ocean Observatory

Data Integration and Visualization

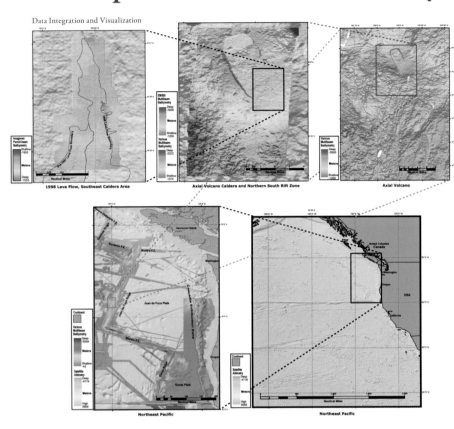

Multibeam Bathymetry of the Northeast Pacific Ocean

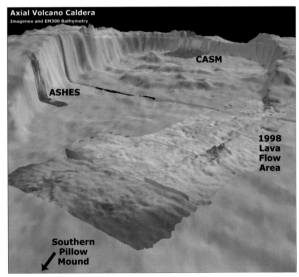

Axial Volcano Caldera
Imagenex and EM300 Bathymetry

CASM

ASHES

1998 Lava Flow Area

Southern Pillow Mound

National Oceanic and Atmospheric Administration, Pacific Marine Environmental Laboratory

Newport, Oregon, USA

By Andra Bobbitt and Susan Merle

Contact

Andra Bobbitt

andra.bobbitt@noaa.gov

Software

ArcInfo 8, ArcView, ArcView Tracking Analyst, Adobe Illustrator, Fledermaus, GMT, and MB system

Hardware

Linux and Windows

Printer

HP Designjet 800

Data Source(s)

National Oceanic and Atmospheric Administration

Natural Resources—Water

Six years of annual visits to the New Millennium Observatory (NeMO) at Axial Volcano have provided the interdisciplinary science team with a large amount of data to manage. Each year a cruise report is compiled including dive logs and maps, sample and experiment information, and discipline summaries. All the cruise-related data is input into ArcGIS. During the five previous field seasons of the NeMO project (summer 1998–2003), this GIS database has been available at sea for use by the scientific party with ArcView.

The added capability of real-time tracking of the remotely operated vehicle with the GIS database using ArcView Tracking Analyst has also been useful. When all the data for the year has been processed (navigation, sample tables) it is brought into several programs for geographical referencing and analysis. Maps can be created using a variety of programs, including ArcGIS, which provides a way to geographically display several types of data at the same time in a legible format. The GIS user can query the database several ways including by samples collected at a site for all years or all instruments deployed and recovered at a site on a particular year.

Managing Louisiana's Public Water Supply With GIS

The U.S. Safe Drinking Water Act amendments of 1996 require all states to develop a Source Water Assessment Program (SWAP) to ensure safe water for all citizens through the protection of water resources. In Louisiana, more than 3,200 public water wells and 50 surface water intakes supply the four and one-half million people of the state along with its estimated 20 million annual visitors.

For the Louisiana SWAP, the only way to fully harness all the components of a vast and diverse public water network in the state was to develop a full-scale GIS. This map is a compilation of some components that represent the many different avenues public water is distributed along with identifying potential risks that could affect quality and/or availability. From 1,000-foot deep wells to intakes along lakes and the Mississippi River, spatial and attribute information was collected so that various calculations could be applied. Geospatial technologies were used to assemble data from more than 20,000 field-collected GPS points, 25 statewide data sets, and information obtained from local water system operators, resulting in an extensive geodatabase.

This information was compiled to assess Louisiana's potable water source sensitivity (using information about the water source including depth, age, groundwater recharge potential, aquifer, and source of water), vulnerability (using distance, ranking, and density calculations of various potential sources of contamination), and susceptibility (a combination of vulnerability and susceptibility). Reports and maps were generated (both in hard copy and via the Web) to notify appropriate state and local individuals of their water source status so that action could be taken if necessary.

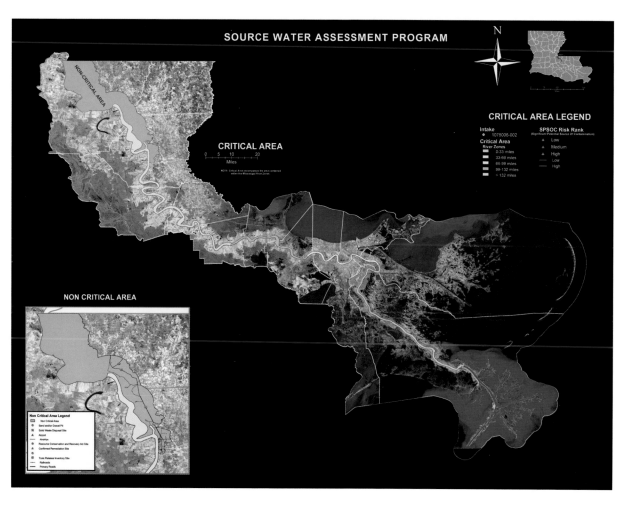

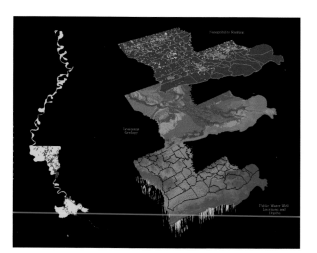

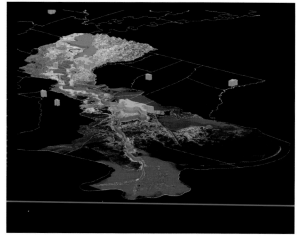

C–K Associates, Geospatial Technologies

Baton Rouge, Louisiana, USA
By John Caldwell, John D. Evers, Perry M. Lopez, and Kevin Moore

Contact

Perry Lopez, C-K Associates
perry.lopez@c-ka.com

John Jennings, Louisiana Department of Environmental Quality
john.jennings@la.gov

Software

ArcGIS 3D Analyst and Microsoft Access

Printer

HP Designjet 5000

Data Source(s)

Field-derived data sets

Natural Resources—Water

Mapping a Diverse Population for Everglades Restoration— A Minority/Low-Income Analysis for Miami-Dade County

Comprehensive Everglades Restoration Plan South Florida Water Management District U.S. Army Corps of Engineers, Jacksonville District

West Palm Beach, Florida, USA

By Heather Kostura and Jerry Krenz

Contact

Heather Kostura

hkostura@sfwmd.gov

Jerry Krenz

jkrenz@sfwmd.gov

Software

ArcMap 8.3

Hardware

Dell Precision Desktop

Printer

HP Designjet 1055cm

Data Source(s)

U.S. Census Bureau 2000 census

Natural Resources—Water

The Comprehensive Everglades Restoration Plan (CERP) will develop numerous water resource/restoration projects throughout south Florida. Because of the rich ethnic diversity in this area, project teams made up of representatives from federal and state agencies must know the demographics of areas potentially impacted by CERP projects. Determining the location of these populations assists the project team when planning outreach materials and activities to ensure maximum public involvement. These maps illustrate the diverse population in the project areas, the neighboring areas, and those areas downstream from restoration projects.

An analysis of the Census 2000 minority and household income data for Miami–Dade County is shown and the legend explains the color block quadrants, which show the relationship between the minority and household income percentages in each census block. The minority and household income thresholds for environmental justice potential areas chosen by the U.S. Environmental Protection Agency Region 4 were used to maintain consistency with the southeast region. In addition to the color coding, the pie charts in each tract represent the percent of Hispanic and Creole speaking populations within that area. This information is invaluable when the project team develops outreach strategies for individuals with limited English proficiency . An inset, not shown, provides a breakdown of minority statistics within the county for quick comparison with other counties.

The complete map set of 16 counties is available to view and download from the Equity Program on the Evergladesplan.org Web site at http://cerp/pm/progr_eee_maps.cfm.

Development Through Knowledge

Human Poverty Indicators

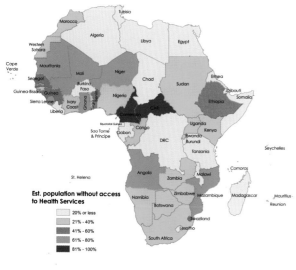

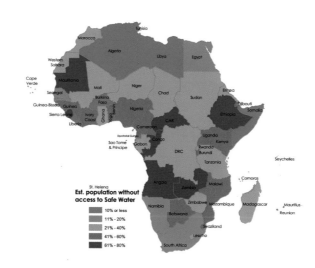

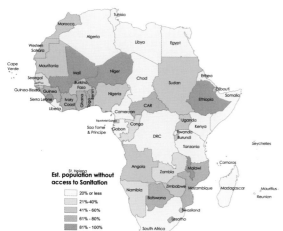

Development Through Knowledge reflects the motto of the Africa Institute of South Africa (AISA), and these maps are a representation of some of the data that AISA publishes in its statistical volume, *Africa at a Glance*. The GIS section of AISA was established in 2003, but it has already made a difference in the way in which AISA presents and disseminates its research outputs and new knowledge. AISA has discovered new ways to talk to its audience by applying the best technology to produce a virtual face of Africa. These maps focus on human development indicators that are important in socioeconomic research.

Adult Literacy

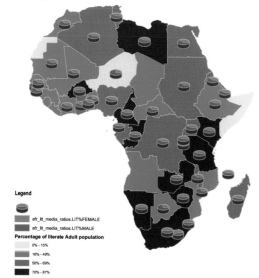

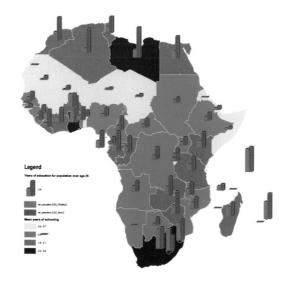

Africa Institute of South Africa

Pretoria, South Africa

By Irene Slabbert

Contact

Irene Slabbert

irene@ai.org.za

Software

ArcView

Printer

HP

Data Source(s)

ESRI and *Africa at a Glance* 2001/2

Sustainable Development

Comprehensive Planning—
Town of Vinalhaven Island, Knox County, Maine

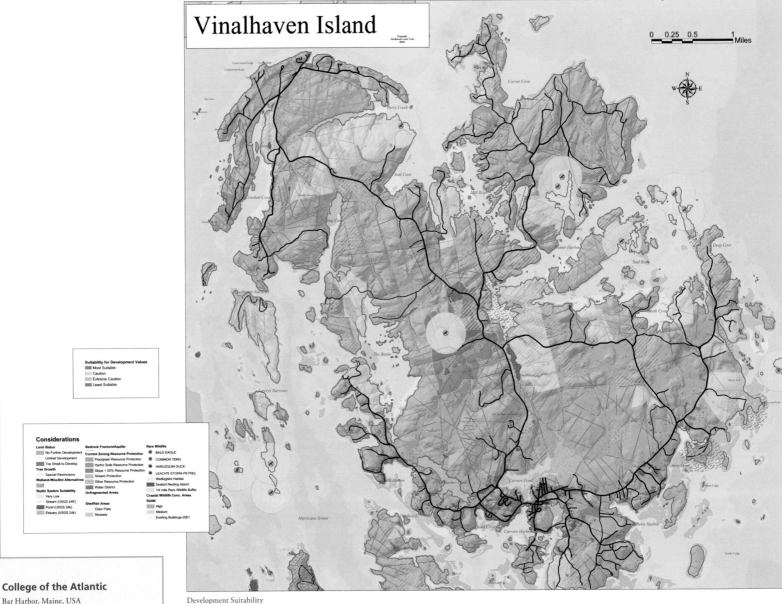

Development Suitability

College of the Atlantic

Bar Harbor, Maine, USA

By Gordon Longsworth and College of the
Atlantic students

Contact

Gordon Longsworth

gordonl@coa.edu

Software

ArcMap 8.3 and Windows 2000

Printer

HP Designjet 650c

Data Source(s)

College of the Atlantic, State of Maine
agencies, U.S. Geological Survey, and
Vinalhaven municipal sources and special
survey sources

Sustainable Development

This display includes images of 12 full-size maps prepared for Vinalhaven Island's Comprehensive Planning process. Geology, soils, surface and groundwater hydrology, land cover, wildlife, marine resources, and land use are represented. The maps provide the town with a comprehensive collection of information, all of which is relevant to planning the future of the island.

The challenge for citizens is not the typical lack of information, but rather how to use all of the information they now have to make the wisest land use decisions. To do this, they rated the most important layers and features for development suitability and assigned colors from red to green with green meaning "Go" and red meaning "Stop." Certain layers, for example, marine resources, were left symbolized while considering potential nonpoint pollution sources of runoff that

might end up in the water. The central Development Suitability map is a summary of values that can be used to guide future development to the most suitable locations. The converse process can be used to help prioritize conservation efforts.

Vinalhaven Island is the largest island of mid-coast Maine, and islanders have been fishing the local waters for generations. They have maintained a sustainable fiishery, one of the most productive for lobster in the state, and want to manage their land in a sustainable way, too. College of the Atlantic has worked closely with the island town to create the maps and assist in the planning process. The town now has GIS capabilities of its own in the high school, town office, and local land trust. It has one of the most comprehensive municipal GIS databases in the state.

Comprehensive Planning—
Town of Vinalhaven Island, Knox County, Maine

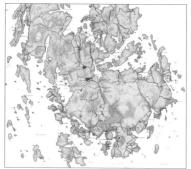

Base Map

Island Parcel Map

Current Zoning Map

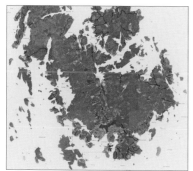

Aerial Photo Map #10

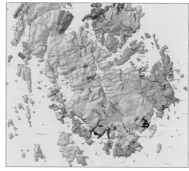

Geology Map

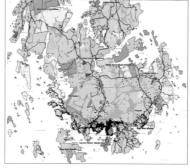

Land Use Map

Soils–Septic System Suitability Map

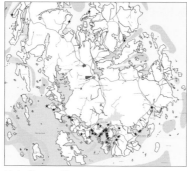

Marine Resources Map

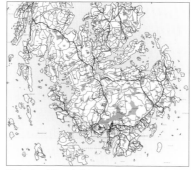

Wetlands and Watersheds Map

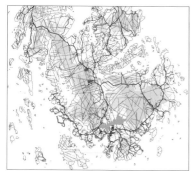

Water Source Index, Watersheds, and Photolinements Map

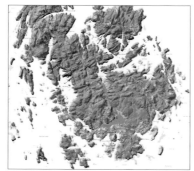

Land Cover Map

Wildlife Map

College of the Atlantic

Bar Harbor, Maine, USA

By Gordon Longsworth and College of the Atlantic students

Contact

Gordon Longsworth

gordonl@coa.edu

Software

ArcMap 8.3 and Windows 2000

Printer

HP Designjet 650c

Data Source(s)

College of the Atlantic, State of Maine agencies, U.S. Geological Survey, and Vinalhaven municipal sources and special survey sources

Sustainable Development

San Diego/Baja California Border Region View From Space

This map displays a false infrared image developed using Landsat 7 imagery remotely sensed during the summer of 2000. The image was created by merging panchromatic (black and white) imagery with selected bands from multispectral imagery to produce a false color infrared composite. The composite image has a 15-meter spatial resolution.

San Diego Association of Governments (SANDAG)

San Diego, California, USA
By John Hofmockel

Contact

John Hofmockel
jho@sandag.org

Software

ArcMap 8.3 and Windows XP

Printer

HP Designjet 800

Data Source(s)

ImageLinks and SANDAG

Sustainable Development

San Diego/Baja California Border Region Planned Land Use

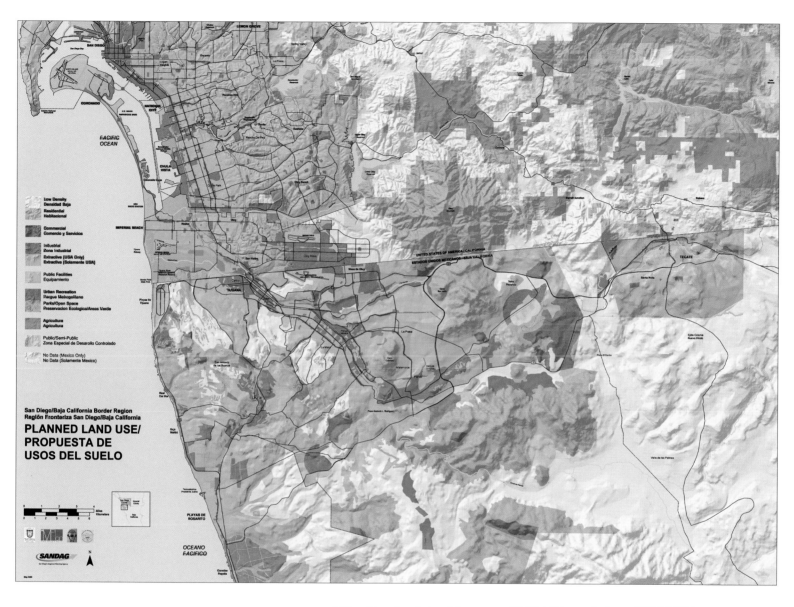

San Diego Association of Governments (SANDAG) has worked closely with local government agencies on both sides of the U.S. Mexican border. The information displayed on this map was compiled from the land use and transportation plans of local jurisdictions in San Diego and Baja California (March 2003). The specific local land use designations were generalized into the categories shown. The freeways and roads shown include proposed roads.

San Diego Association of Governments

San Diego, California, USA

By John Hofmockel

Contact

John Hofmockel

jho@sandag.org

Software

ArcMap 8.3 and Windows XP

Printer

HP Designjet 800

Data Source(s)

H. Ayuntamiento Tijuana, Tecate, Instituto Municipal de Planeación, Playas de Rosarito, SANDAG, SanGIS, and U.S. Geological Survey

---

Sustainable Development

Colombia, Mapa Guia 2003

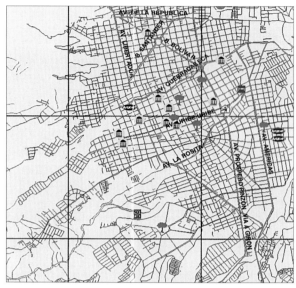

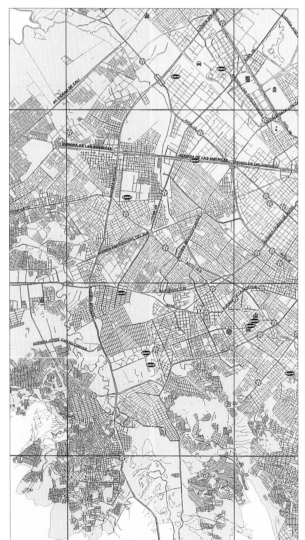

Geodigital Ltda.
Bogotá, Colombia
By Geodigital, Ltda.

Contact
Gustavo Montoya
geodigital@tutopia.com,
montgus@hotmail.com

Software
ArcInfo, ArcView 3.2, CorelDRAW, and
ERDAS IMAGINE

Hardware
PC Pentium 4 and Power Mac G4

Printer
Off-set printer

Data Source(s)
ESRI, Dane–Colombia, Geodigital,
and Invias

Tourism

Colombia, Mapa Guia is a two-sided tourist road map (90 by 60 centimeters). One side shows Colombia's political divisions, roads, points of interest, towns and cities larger than 12,000 inhabitants, natural parks, and a contrasted digital elevation model, which clearly conveys the importance of Colombia's mountainous terrains. It contains an indexed list of the most important towns in the country and a distance table for major cities.

On the opposite side, 11 of the most important Colombian cities are featured with their main roads and points of interest highlighted.

D.C. Vicinity Map and Visitor's Guide

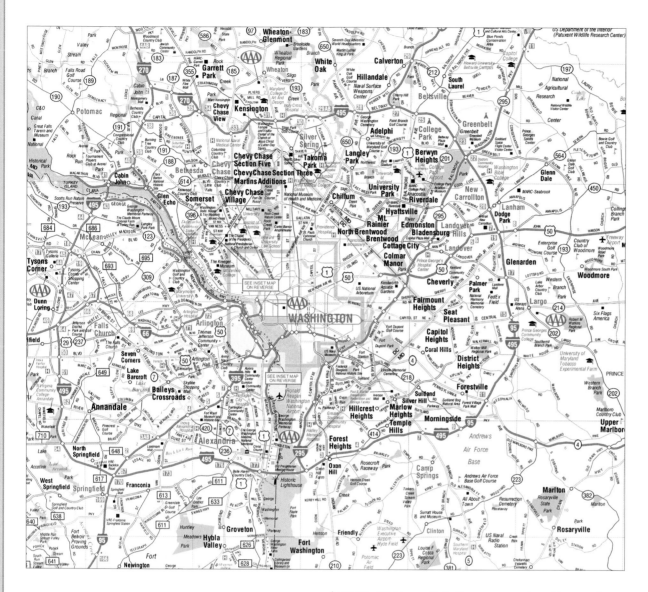

The Washington, D.C., Vicinity map was derived from AAA's detailed nationwide ArcSDE database. All AAA GIS maps, whether in digital or paper format, contain proprietary information provided by AAA's various travel information departments and are designed to meet the travel needs of its members. GIS cartographers generalize the spatial data for each map product using rules for specific map types and scales. The AAA GIS ArcSDE master database contains more than 50 GB of data and supports cartographic efforts in both paper and electronic applications. Cartographic edits are applied to product specific layers, while generalized edits are applied to the master data layers. Individual map views have custom projection parameters. The Washington, D.C., Vicinity map illustrates the data presentation flexibility inherent with the scalable data within the AAA GIS database.

AAA has been producing quality maps since 1911. AAA GIS incorporates traditional cartographic techniques, automated production processes, and a multilayered GIS database to produce and update annually more than 65 sheet maps, 600 TourBook maps, and a North American Road Atlas for the association's 47 million members.

AAA
Heathrow, Florida, USA
By AAA GIS

Contact
Michael Mouser
mmouser@national.aaa.com

Software
ArcInfo 8.1 and ArcSDE

Printer
Epson Stylus Pro 9000

Data Source(s)
AAA and Navteq

Tourism

Local Area Map for London Underground Stations

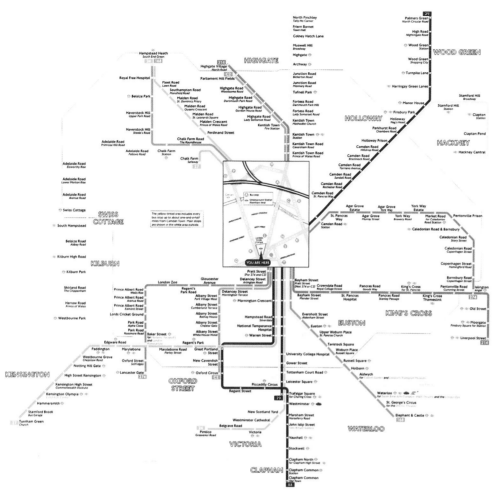

Transport for London
London, United Kingdom

T-Kartor Sweden AB
Kristianstad, Sweden
By T-Kartor Sweden AB

Contact
David Figueroa
df@t-kartor.se

Software
ArcMap, Maplex 3.4, Adobe Illustrator and
InDesign, and CPS NG

Printer
Printed using large-format digital technology

Data Source(s)
Collins Bartholomew and Transport for
London

Tourism

G etting around London can be a challenge. Each day approximately two million customers use the underground service alone. In addition, there are hundreds of thousands who ride buses, trains, and other transport services covering the London area. Many of these people know exactly where they are going, but there are many who do not. Aiming to make the door-to-door journey for all underground passengers as easy as possible, Transport for London has created a Local Information Display for all 275 London underground stations consisting of a local area map and a bus service "Spider Map" informing customers of nearby places reachable from the station at street level.

The local area map contains detailed place of interest information collected in large part from underground station staff who respond to hundreds of questions daily regarding nearby places and attractions. The result is a large poster display located conveniently near the station exits. In addition, to save customers from scrambling about trying to find pen and paper, A4 format leaflets have been printed containing a condensed version of the information on the poster for free distribution.

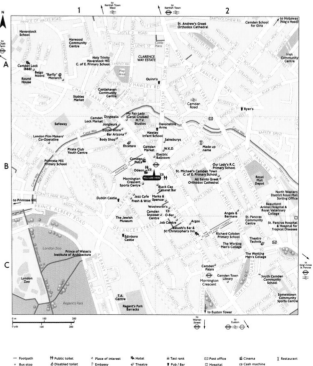

Trail of the Coeur d'Alenes

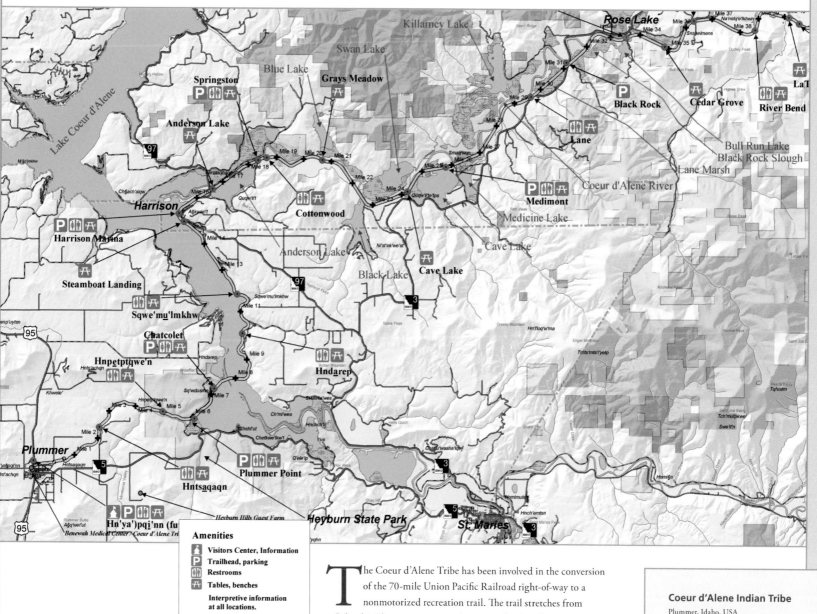

Amenities

🏛️ Visitors Center, Information

🅿️ Trailhead, parking

🚻 Restrooms

🪑 Tables, benches

Interpretive information at all locations.

Legend

◉ Trail of the Coeur d'Alenes Amenities
✛ Mile Markers
● Sponsor Locations
═══ Trail of the Coeur d'Alenes
--- Coeur d'Alene Reservation Boundary
── Primary Roads
── Secondary Roads
── Gravel Roads
── Streams
Lakes, Rivers
Marsh
Marsh, submerged
BLM
USFS
State of Idaho
Hn'l'l'oq'w'ma : Coeur d'Alene Place Names

The Coeur d'Alene Tribe has been involved in the conversion of the 70-mile Union Pacific Railroad right-of-way to a nonmotorized recreation trail. The trail stretches from Columbia Plateau country on the western border of the panhandle of Idaho to the Bitterroot Mountains on the Idaho–Montana border. The trail traverses rolling farmlands, cedar and pine forests, spectacular wetlands along the Coeur d'Alene River, and the historic Silver Valley mining district.

The purpose of creating the map was to provide a guide to the new trail and enable its users to navigate it more easily. It also demonstrated the ability of the Tribe's GIS group to produce quality commercial maps for the public.

Coeur d'Alene Indian Tribe

Plummer, Idaho, USA

By John Hartman and Berne Jackson

Contact

Berne Jackson

bjackson@cdatribe-nsn.gov

Software

ArcMap, ArcSDE, Adobe Photoshop, and Windows 2000

Hardware

Pentium III workstations and Ultra 160 SCSI drives

Printer

HP Designjet 5500

Data Source(s)

Coeur d'Alene Tribal trails and other base ArcSDE data

Tourism

The Heart of Kiger Gorge, Oregon

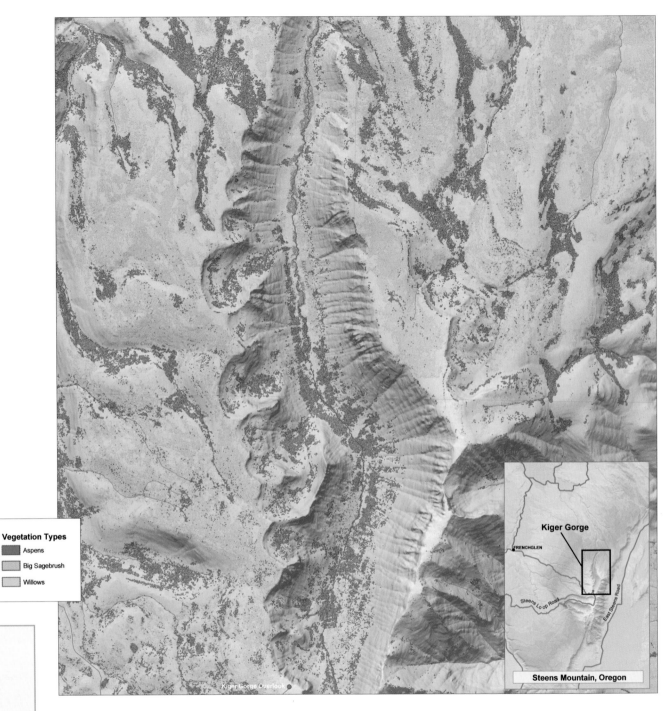

Elevations

High : 9468 Feet

Low : 4604 Feet

Vegetation Types

- Aspens
- Big Sagebrush
- Willows

Steens Mountain, Oregon

Kiger Gorge

FRENCHGLEN

Steens Loop Road

East Steens Road

U.S. Department of the Interior, Bureau of Land Management

Portland, Oregon, USA

By Kelly Hazen and Jeffery S. Nighbert

Contact

Jeffery Nighbert

jnighber@or.blm.gov

Software

ArcGIS 8.3 and ArcGIS Spatial Analyst

Printer

HP Designjet 1055cm

Data Source(s)

U.S. Geological Survey

Tourism

Kiger Gorge is one of five deep canyons that cut through Steens Mountain located in southeastern Oregon. The heart of Kiger Gorge illustrated in this map typifies the dramatic nature of these gorges, which were created by mountain glaciation and other geologic events during the last ice age. More than 2,000 feet deep, Kiger Gorge offers a unique opportunity for experiencing high desert solitude and breathtaking views of dramatic landscapes. Exploration is limited to hiking or horseback riding because there are no formal trails.

This poster map was created for the Bureau of Land Management's Burns District Office in Burns, Oregon, to illustrate and advertise the beauty of Kiger Gorge, which occupies the center of the Steens Mountain complex in southeastern Oregon. The map will also be used as a hiking map as well as a poster. On the technical level, this map uses new "bump mapping" techniques, which create realistic textures representing different land cover types. These textures are generated using ArcGIS Spatial Analyst commands and attempt to characterize through shadow and color the look and feel of different vegetation types.

For more information, go to www.or.blm.gov/Burns.

Building Tomorrow's Data Warehouse Today

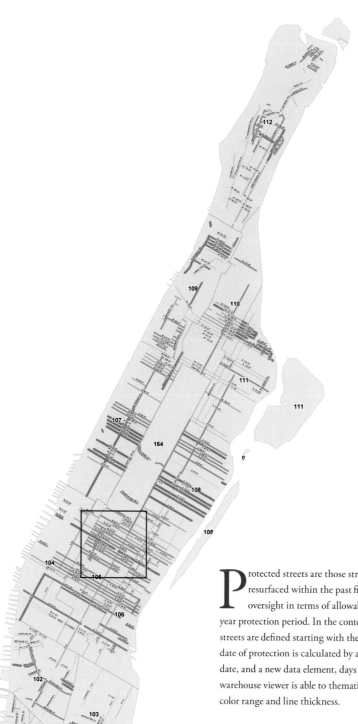

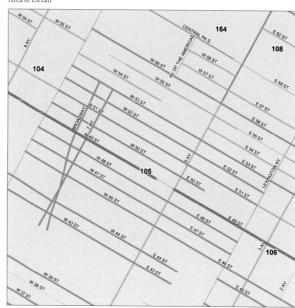

Area of Detail

P rotected streets are those streets that have been reconstructed or resurfaced within the past five years and are subject to greater oversight in terms of allowable work on the street during the five-year protection period. In the context of the data warehouse, protected streets are defined starting with the capital project completion date. The date of protection is calculated by adding 60 months to the completion date, and a new data element, days to expiration, is created. The data warehouse viewer is able to thematically map the street segments using color range and line thickness.

By highlighting the streets soon to expire from protection, planners can better schedule work in the streets so that roadbeds are cut less frequently and, at times, just before they are likely to come up for scheduled resurfacing work.

Bowne Management Systems, Inc.

Mineola, New York, USA

By Robert Nossa

Contact

Robert Nossa

rnossa@bownegroup.com

John Griffin

jgriffin@dot.nyc.gov

Software

ArcGIS 8.2 and Windows 2000

Hardware

Dell Inspiron 8000

Printer

HP Designjet 5500

Data Source(s)

New York City

Transportation

Public Transport Optimization in the Liberec Region Project

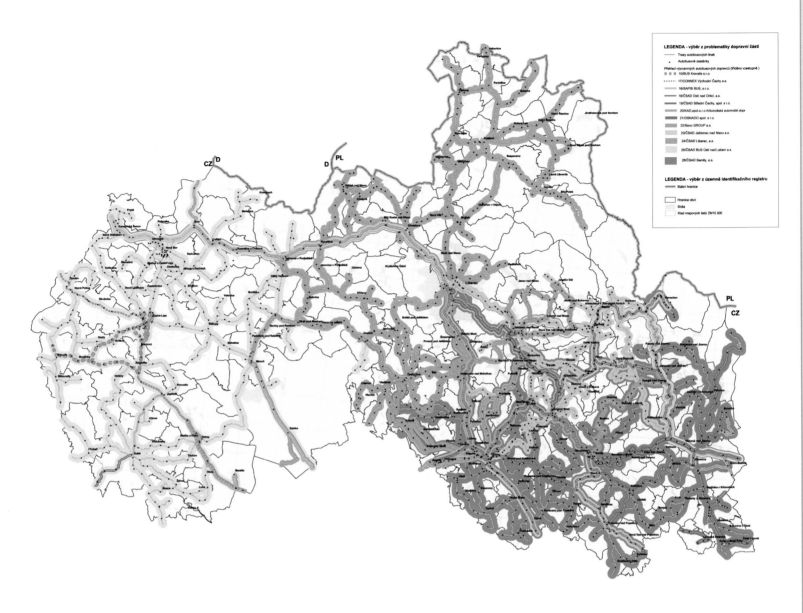

Liberec Region

Liberec, Czech Republic

By Kamila Mikulecka in cooperation with
EMA spol. s.r.o.

Contact

Kamila Mikulecka
kamila.mikulecka@kraj-lbc.cz

Software

ArcInfo 8.3

Hardware

Dell Precision 340

Printer

HP Designjet 800ps

Data Source(s)

Liberec Region GIS

Transportation

Public Transport Optimization in the Liberec Region Project evaluates current railway and bus transportation systems. Optimization should contribute to higher efficiency of present public transport services with respect to the connection of neighboring regions and future implementation of an integrated transportation system.

The transport system analysis consists of the creation of digital geographic data—information modules such as timetables and number of people transported. The map shows bus stops and bus routes in the Liberec Region. Based on the data collected from transport companies during January and February 2003, a registry of transported persons was created. There are great differences in the use of public transportation in various part of Liberec Region depending on the population density.

This presentation consists of four maps. The first demonstrates the quantities of transported persons in different areas of the Liberec Region in a specified time. The second map contains routes classified by transport company. The third map shows the number of persons on each bus stop in the specified time, and the fourth map evaluates the use of transport segments between bus or railway stations in the specified time.

Public Transport Optimization in the Liberec Region Project

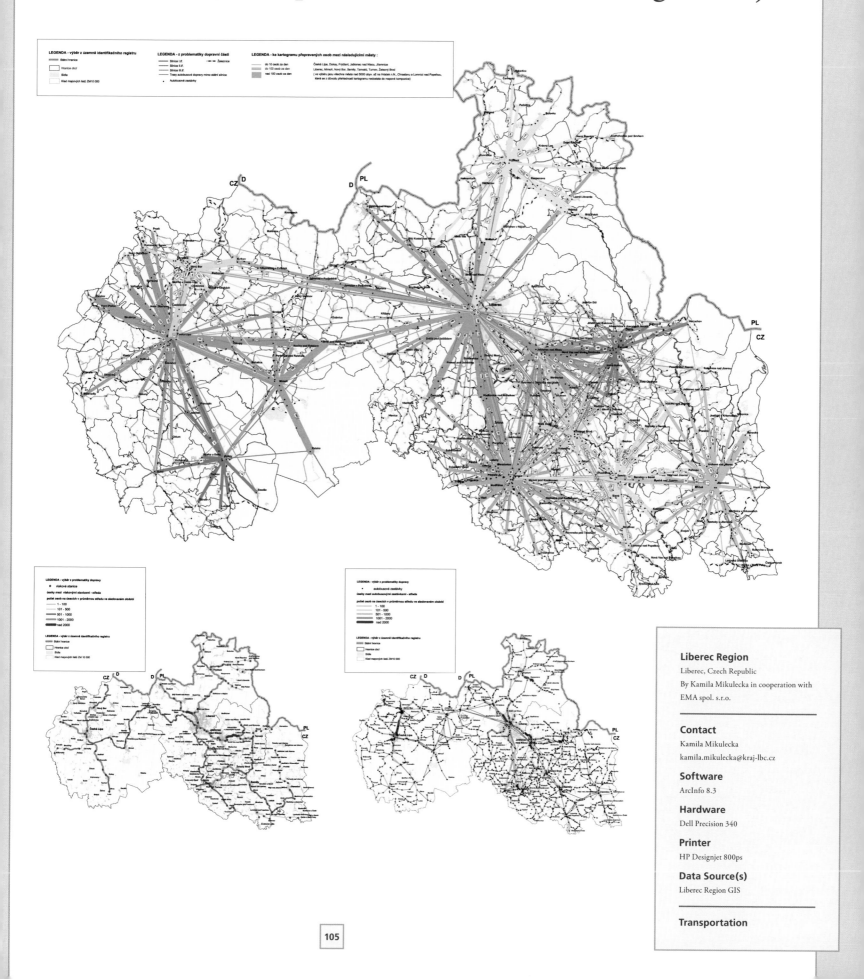

Liberec Region

Liberec, Czech Republic

By Kamila Mikulecka in cooperation with
EMA spol. s.r.o.

Contact

Kamila Mikulecka
kamila.mikulecka@kraj-lbc.cz

Software

ArcInfo 8.3

Hardware

Dell Precision 340

Printer

HP Designjet 800ps

Data Source(s)

Liberec Region GIS

Transportation

Railroad Through 150 Years in Norway

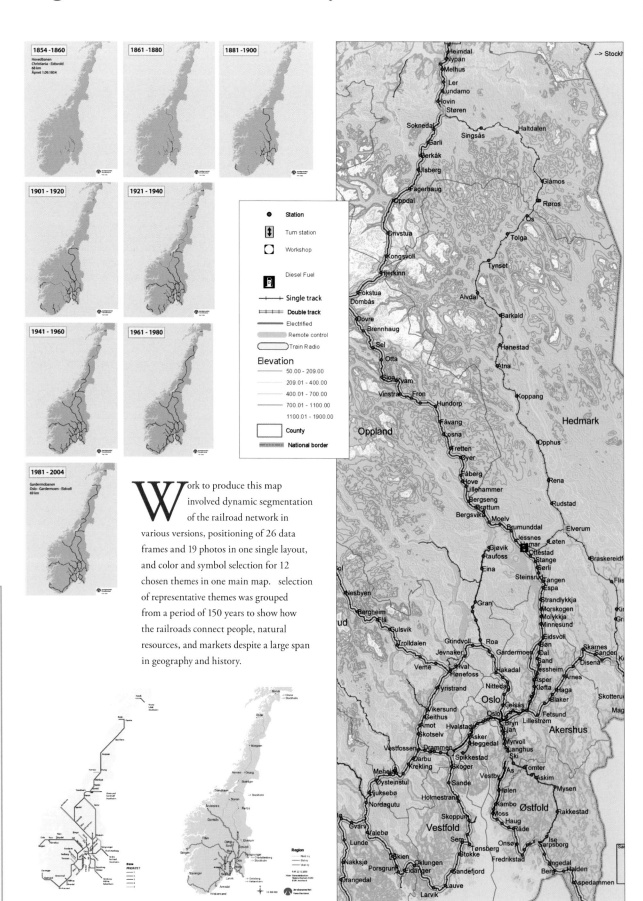

The Norwegian National Rail Administration

Oslo, Norway

By Per Anton Fevang

Contact

Per Anton Fevang

paf@jbv.no

Software

ArcView 8.3

Hardware

HP Compaq Pentium 4

Printer

HP Designjet 1050c+

Data Source(s)

Network Statement 2005, Norwegian Map Authorities Network, Norwegian Railway Database, Railway Data 1994, and Railway Track Network

---

Transportation

W ork to produce this map involved dynamic segmentation of the railroad network in various versions, positioning of 26 data frames and 19 photos in one single layout, and color and symbol selection for 12 chosen themes in one main map.   selection of representative themes was grouped from a period of 150 years to show how the railroads connect people, natural resources, and markets despite a large span in geography and history.

Interstate 680 North Corridor Concept

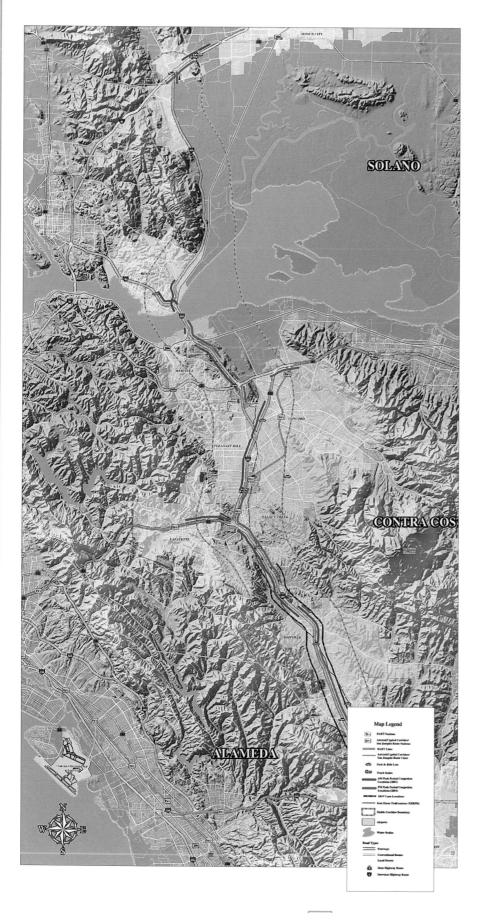

California Department of Transportation (Caltrans) has the statutory responsibility to engage in regional transportation planning, which includes the development of transportation corridor concept reports and other planning activities in partnership with other local planning agencies. Caltrans' goal is a safe, sustainable transportation system that is environmentally sound, socially equitable, economically viable, and developed through collaboration. It provides for the mobility and accessibility of people, goods, services, and information through an integrated multimodal network.

These planning efforts require significant amounts of research and analysis before conclusions can be reached. Various efforts are underway at the state, regional, and local levels to identify appropriate performance measures to evaluate overall system efficiency and for the prioritization of specific transportation improvements. District 4 is developing a methodology to identify the ideal sequence of highway projects that should be implemented based on knowledge of existing bottlenecks, congestion, and intermodal connectivity needs.

California Department of Transportation

Sacramento, California, USA

By Matthew Kelly

Contact

Monica Markel

monica.markel@dot.ca.gov

Software

ArcGIS Spatial Analyst, Adobe Illustrator, and Windows 2000 Professional

Printer

HP Designjet 5000ps

Data Source(s)

Caltrans and U.S. Geological Survey

Transportation

Transit Commuting in America, 1990 to 2000

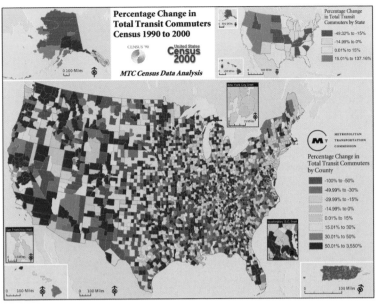

U.S. Counties

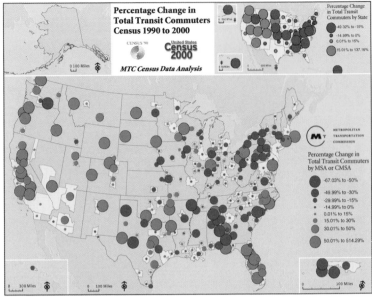

U.S. MSAs and CMSAs

36 Largest U.S. MPOs

Metropolitan Transportation Commission

Oakland, California, USA

By Chuck Purvis, Mike Skowronek, and
Garlynn G. Woodsong

Contact

Garlynn Woodsong

gwoodsong@mtc.ca.gov

Software

ArcGIS 8.2, Quark Express 5.0, and
Windows 2000

Hardware

Gateway Pentium III

Printer

HP Designjet 5000ps

Data Source(s)

U.S. Census Bureau (Summary File #3,
TIGER/Line 2000)

Transportation

This set of maps is based on journey to work data from the 1990 and 2000 decennial U.S. censuses. Data is derived from the long form census questionnaire. One in six U.S. residents filled out the census long form. Transit commuting includes bus or trolleybus, streetcar or trolley, subway or elevated rail, railroad, and ferryboat (excludes taxicab).

Data is mapped at the county, metropolitan, largest metropolitan planning organization (MPO), region, and state levels and includes information for 3,219 counties, 280 metropolitan statistical areas (MSAs) and consolidated metropolitan statistical areas (CMSAs), the 36 largest MPO regions (MPOs with a population greater than one million), and 52 states/state equivalents (includes the District of Columbia and Puerto Rico).

The purpose of these maps is to show changes in transit commuting within the United States between 1990 and 2000 at various geographic levels. The maps are useful in portraying the increase in transit commuting in "sunbelt" metropolitan areas contrasting with the decrease in transit commuting in "rustbelt" metro areas. These maps were also used to support a Transportation Research Board presentation entitled "Transit Ridership and Transit Commuting Trends: Are They Different?" It discussed the reasons for the census showing a 0.4 percent decrease in transit commuting between 1990 and 2000, compared to a 6.4 percent overall increase in transit ridership over this same period.

Washoe County Environmental Justice— Population and Race by Census Block

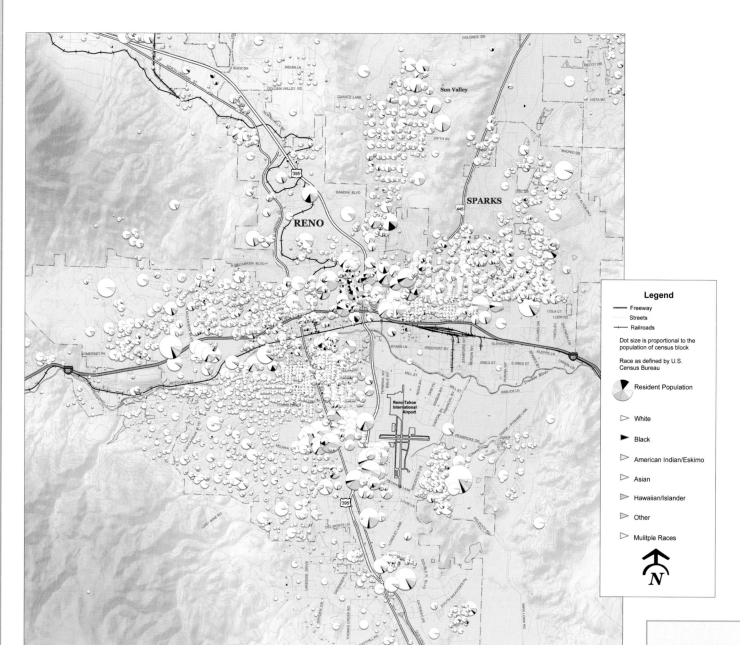

Legend

— Freeway
— Streets
┼ Railroads

Dot size is proportional to the population of census block

Race as defined by U.S. Census Bureau

Resident Population

▷ White

▶ Black

▷ American Indian/Eskimo

▷ Asian

▷ Hawaiian/Islander

▷ Other

▷ Mulitple Races

N

Parsons Corp. is looking at the demographics of Washoe County (Reno–Sparks, Nevada) to find a route for a beltway around the urban core of Reno and Sparks. Factoring in forecasted residential and employment growth, GIS helps to find suitable routes that serve the greatest number of people now and will accommodate the expectant growth during the next 20 years.

With several alternative routes available, census data is plotted on a map of the county to show the racial demographics in the study area. With this map, planners can choose the route that does not unfairly impact minority populations. Similar maps help planners avoid impacts to low-income populations.

This map shows the population distribution by census block for Washoe County. Each census block is represented by a dot, with a size relative to the total population of that census block. Each dot is actually a pie chart, representing the population by race (as defined by the U.S. Census Bureau).

Parsons Corp.

San Jose, California, USA

By Eric Coumou

Contact

Eric Coumou

eric.coumou@parsons.com

Software

ArcInfo 8.2 and Windows NT

Printer

HP Designjet 1050c+

Data Source(s)

U.S. Census Bureau and Washoe County Planning Department

Transportation

World Air Traffic Flowchart

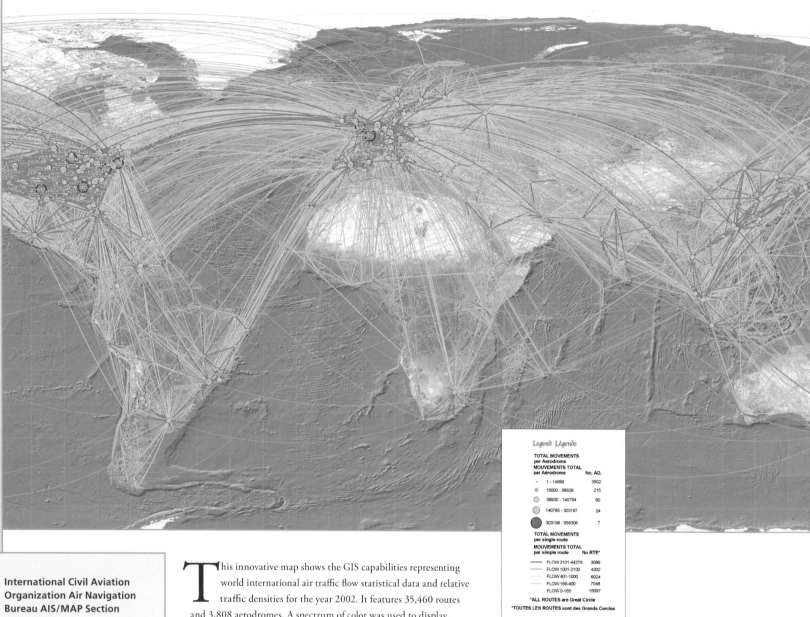

**International Civil Aviation
Organization Air Navigation
Bureau AIS/MAP Section**

Montréal, Québec, Canada

By Gilbert Lasnier

Contact

Gilbert Lasnier

glasnier@icao.int

Software

ArcIMS, ArcView 8.3, and Microsoft Access

Hardware

Compaq Evo W4000

Printer

HP Designjet 1050c

Data Source(s)

International Civil Aviation Organization
and WorldSat International, Inc.

Transportation

This innovative map shows the GIS capabilities representing world international air traffic flow statistical data and relative traffic densities for the year 2002. It features 35,460 routes and 3,808 aerodromes. A spectrum of color was used to display overlapping routes, ranging from blue to red, which when viewed with ChromaDepth™ glasses will separate three-dimensionally into their respective layers.

Aerodromes are presented as relative traffic-graded symbols by which total movements for each aerodrome are displayed. Routes were calculated using a great circle calculation, which rendered the shortest distance between destinations. Some 350,000 great circle vectors have been included in this chart.

The aerodromes were featured as relative-valued graduated symbols, which were modified many times. Total movements for each aerodrome were featured as a subtotal calculated with reference to the spatial analysis. All routes were featured as great circles calculated using a great circle calculation that obtained the shortest distance between two destinations.

Creating this map was a challenging exercise. Demonstrating all the data on the map clearly and displaying areas with overlapping routes using color spectrum ranging from blue to red were difficult tasks. The difficulties encountered in this exercise were the number of vectors that needed to be generated in Visual Basic, records that neared zero degree gave false results, and the number of vector lines that passed across the longitude at 180 degrees east had to be managed differently to be reconnected at 180 degrees west longitude.

This data will be loaded in an ArcIMS application to enable users to interactively navigate and query a range of information. A reference tool is under development to address the need for determining more efficient route systems and analyzing the most cost-effective routes through flight information regions.

Electric Utility Maps

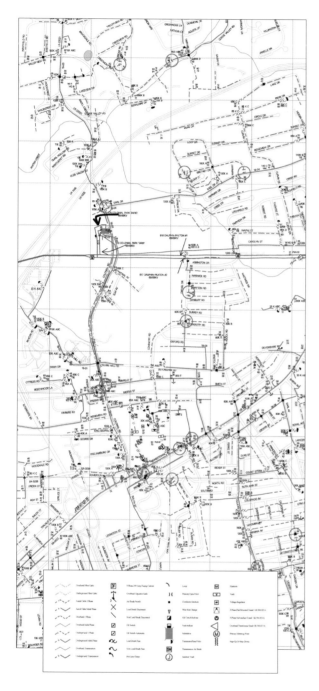

Storm Wall Map

The Storm Wall Map 240-348-S8 shows the PPL transmission (purple), substations (green), and distribution (blue) facilities on the PPL land base (gray/blue water) at 1:800 scale. Primary sectionalizing device symbols are colored red, and normally open tie device symbols are colored red and circled in red. Storm Wall maps are used by PPL for tracking storm outages and repair crews working in PPL's 12 operating area offices as it manages the restoration efforts following a weather event. All storm maps for an area are mounted on a wall in an emergency room and covered with Plexiglas so both outages and repair crew locations can be plotted until the restoration effort is completed.

Primary Operating Map

The PPL Primary Operating Map (POM) 252-348-S4 shows the PPL 12 kilovolt distribution system. PPL POMs are used by PPL for daily operation of the overhead distribution system. These maps are updated as new facilities are added using Miner & Miner's JP/DE engineering application, or corrections are edited using the MMEdit application. Copies of all POMs for each of PPL's 12 operating areas are carried in map containers in all the construction/repair trucks in each area. The maps on the vehicles are replaced on a predetermined cycle. These maps are used in the field primarily for finding the location of PPL facilities and by repair crews and damage assessment personnel during outage events.

PPL Electric Utilities

Allentown, Pennsylvania, USA

By PPL Electric Facilities Database Team

Contact

Sheldon Seip

ssseip@pplweb.com

Software

ArcInfo 7.2.3, Miner & Miner MMView

Hardware

HP v2250

Printer

HP Designjet 1055cm+

Data Source(s)

Electric facilities database

Utilities—Electric and Gas

Kárahnjúkar Hydroelectric Project, Hálslón Reservoir

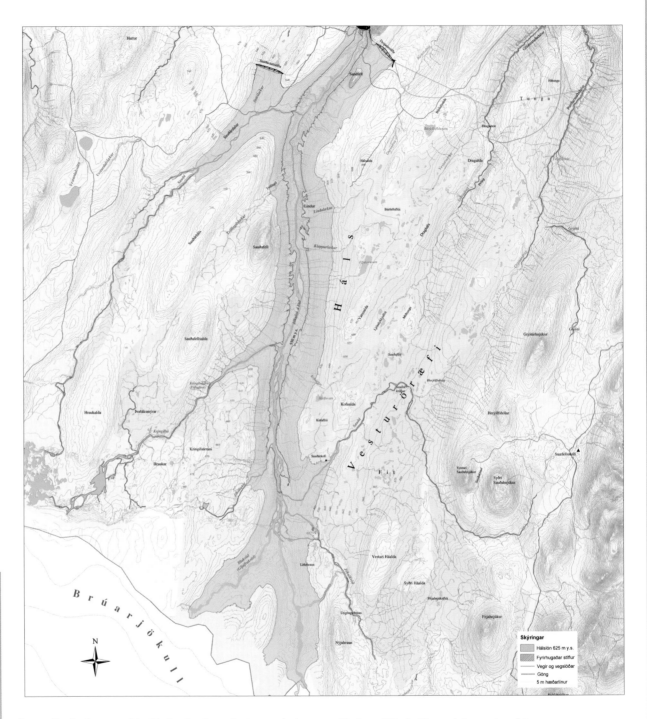

The National Power Company of Iceland

Reykjavik, Iceland

By Theodór Theodórsson

Contact

Theodór Theodórsson

teddi@lv.is

Software

ArcMap 8.3 and ArcView

Hardware

Dell Optiplex GX110

Printer

HP Designjet 2500cp

Data Source(s)

Icelandic Institute of Natural History, National Energy Authority, National Land Survey of Iceland and Public Roads Administration, and custom CAD design data

---

Utilities—Electric and Gas

The development of the Kárahnjúkar Power Station entails the harnessing of the glacial rivers Jökulsá á Dal and Jökulsá í Fljótsdal in northeastern Iceland. These rivers originate in the northeastern region of the Vatnajökull ice cap and run in a northeast direction through the Jökuldalur and Fljótsdalur valleys to their common estuary in the Héradsflói Bay.

The hydropower station will have an installed capacity of 630 megawatts, a harnessed flow rate of 126 cubic meters per second, and a power generating capacity of 4,450 gigawatt hours per year. The hydropower project involves the damming of the glacial river Jökulsá á Dal at Mt. Fremri Kárahnjúkar and the creation of the water storage reservoir Hálslón. Another small reservoir, Ufsarlón, will be created by damming the glacial river Jökulsá í Fljótsdal.

From the Hálslón reservoir, water is conveyed through an underground head race tunnel eastward and joins another tunnel from the Ufsarlón reservoir. The water is then carried in a single tunnel northeastward to the Teigsbjarg escarpment, where it drops through two steep penstocks to an underground powerhouse. There the water enters six generating units in the powerhouse and travels through a tailrace tunnel and canal into the course of the glacial river Jökulsá í Fljótsdal.

Electric Facilities

Legend

S	Substation
C	Circuit Breaker
M	Metering Point
	PME-4
	PME-6
	PME-9
	PME-9A
	PME-11
	Disconnect Switch
AS	Air Break Switch
LAS	Load Break Air Switch
SDC	Sectionalizing Disconnect Switch
	Load Break Elbow
	Load Break Switch
	1 Phase Cabinet
	3 Phase Cabinet
	Recloser
S	Sectionalizer
	Open Point
H	Handhole
	Manhole
VU	Vault
	Capacitor
	Voltage Regulator
	OH Fuse, ABC
	OH Fuse, C
	OH Fuse, B
	OH Fuse, BC
	OH Fuse, A
	OH Fuse, AC
	UG Fuse, ABC
	UG Fuse, C
	UG Fuse, B
	UG Fuse, A

	Unk Transformer
	UG Transformer, ABC
	Unk UG Transformer
	UG Transformer, A
	UG Transformer, B
	UG Transformer, C
	OH Transformer, ABC
	Unk OH Transformer
	OH Transformer, A
	OH Transformer, B
	OH Transformer, C
	UM-8-3 Pedestal
	UK-5 Pedestal
	UK-6 Pedestal
	Wood Pole
	Non-wood Pole
H	H-frame Pole
	Unknown Misc Feature
	Riser
	Aerial Marker
	Fault Indicator
	Fault Limiter
	Ground
	Splice
	Surge Arrester
	Warning Sign
	Bulk Delivery Point
MP	Primary Meter
	Service Point
	Security Light
	Street Light
	Down Guy
	Unk Primary OH
	Substation Bus Bar, ABC
	3 Phase Primary OH, ABC
	2 Phase Primary OH, AC
	2 Phase Primary OH, BC
	2 Phase Primary OH, C
	1 Phase Primary OH, C
	1 Phase Primary OH, B
	1 Phase Primary OH, A
	Unk Primary UG
	Switch Cabinet Bus Bar, ABC
	Switch Cabinet Bus Bar, A
	Switch Cabinet Bus Bar, B
	Switch Cabinet Bus Bar, C
	3 Phase Primary UG, ABC
	2 Phase Primary UG, AC
	1 Phase Primary UG, A
	1 Phase Primary UG, B
	1 Phase Primary UG, C
	Span Guy
	Unk Secondary OH
	1 Phase Secondary OH
	3 Phase Secondary OH
	OH Streetlight Conductor
	Unk Secondary UG
	1 Phase Secondary UG
	3 Phase Secondary UG
	UG Streetlight Conductor

This map is a document used by the CoServ GIS department to provide electric crews and engineers with information in a more consistent, uniform manner. Standardized rendering and symbology enable CoServ to communicate with the electric crews more effectively. The use of color in this map is important because red, green, and blue not only mean something to the crew, but these colors are also meaningful in the field. Electric crews and engineers can use the map to identify lines and equipment by phase (e.g., single phase A, single phase B, single phase C, or A phase transformer). Complex rendering can aid engineers in electric load balancing, line crews in construction and switching, and dispatchers in outage situations.

Map symbology will continue to play an integral part in the way GIS communicates with those who are in the field. As more and more data fields are attributed, the GIS team at CoServ can dig deeper into exploring new ways to communicate through symbology and labeling.

CoServ Electric

Corinth, Texas, USA

By Michael Walden

Contact

Michael Walden

mwalden@coserv.com

Software

ArcFM™ 8.3, ArcGIS 8.3, ArcSDE, Oracle9i, and Windows XP Professional

Hardware

Dell Laptop Pentium 4

Printer

HP Designjet 1055cm+

Utilities—Electric and Gas

Hurricane Isabel—A Storm of Mass Destruction and an Unprecedented Restoration Effort for Dominion

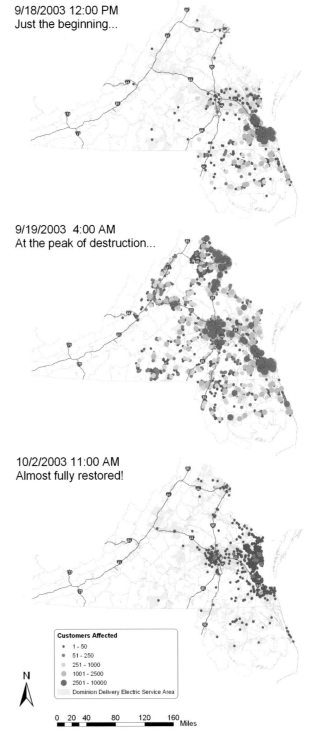

9/18/2003 12:00 PM
Just the beginning...

9/19/2003 4:00 AM
At the peak of destruction...

10/2/2003 11:00 AM
Almost fully restored!

Customers Affected
- 1 - 50
- 51 - 250
- 251 - 1000
- 1001 - 2500
- 2501 - 10000

Dominion Delivery Electric Service Area

N

0 20 40   80   120   160 Miles

Dominion
Richmond, Virginia, USA
By David L. Reed

Contact
David L. Reed
david_l._reed@dom.com

Software
ArcGIS 8.3 and Windows 2000

Hardware
Dell Latitude/Intel Pentium III

Printer
HP Designjet 1055cm+

Data Source(s)
Dominion GIS department

Utilities—Electric and Gas

Dominion is one of the nation's largest producers of energy and serves approximately 2.2 million regulated electric power retail customers through more than 60,000 miles of distribution systems in Virginia and northeastern North Carolina.

Hurricane Isabel was the worst natural disaster to hit the Dominion electric service territory. The weather event disrupted electric service to 82 percent of the 2.2 million customers across the 30,000-square-mile service territory. GIS played a critical role in the restoration effort. The technology was used to assess damage, dispatch crews and, most important, communicate power restoration progress to customers.

As the restoration effort ended, a need for a map depicting the magnitude of the destruction of the storm from a chronological perspective became apparent. This map shows outages that were processed by the outage management system during the storm and the days after the storm. A timeline also tells the story of unprecedented efficiency and cooperation to restore power to customers.

Electric System Integration at the Truckee Donner Public Utility District

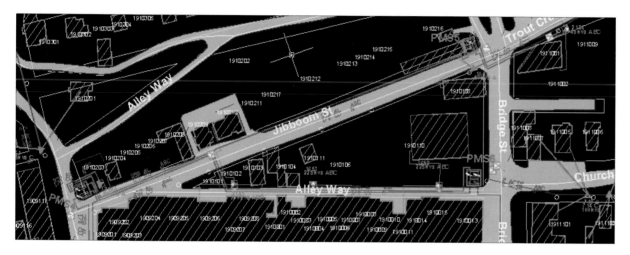

The Truckee Donner Public Utility District (PUD) provides electric and water service to the town of Truckee. Located in the central Sierra Nevada Mountains, Truckee currently has approximately 15,000 permanent residents but is experiencing rapid growth. Historically, PUD has stored spatial data in individual legacy files and decaying Mylar and paper plots. Three years ago, PUD started the process of converting legacy files into a central geodatabase, integrating spatial and attribute information from several divisions of the enterprise. The geodatabase model provides a central, versioned, spatial data repository; the flexibility to integrate with other platforms; and advanced modeling capabilities.

The GIS department has become an integral hub of the PUD organization. GIS has allowed PUD to integrate customer information, SCADA information, and design specifications with facilities and land base data.

The PUD accomplishes pole analysis using a combination of ESRI GIS, Miner & Miner structural analysis tools, and a custom utility for viewing phase designation on overhead lines that was developed by PUD and POWER Engineers. The GIS stores attribute information from pole inventories and digital photos of all poles, which can be viewed using hyperlinks.

Secondary circuit analysis is accomplished with the ArcFM suite. Analysis tools enable district engineering staff determine the optimal transformer size based on the number of customers served by that transformer. Secondary analysis also helps the engineering staff determine transformers that could be overburdened.

The ArcGIS Schematics extension uses electrical network data to generate online schematic diagrams of electric circuits. Schematic diagrams are useful to engineering staff in system design and to operations staff who use schematic diagrams to determine switching scenarios during outage situations.

Integrating CAD into the geodatabase model has facilitated communication between district staff and the engineering community. Staff can also use existing CAD data and has developed an export utility that quickly and easily exports GIS data to a CAD format.

Truckee Donner Public Utility District
Truckee, California, USA
By Steve Murphy

Contact
Steve Murphy
stevemurphy@tdpud.org

Software
ArcFM, ArcGIS Schematics, ArcMap, Advantica Stoner SynerGEE, AutoCAD, Custom GIS applications developed by POWER Engineers, and Windows

Hardware
Dell workstation

Printer
HP Designjet 1055c

Data Source(s)
Aerial photography, CAD, digital photography, Metroscan, and proprietary facilities data

Utilities—Electric and Gas

Water Utility Modeling at the Truckee Donner Public Utility District

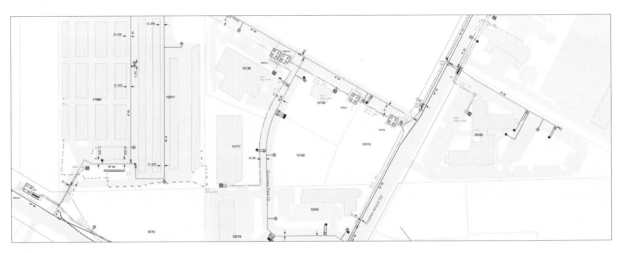

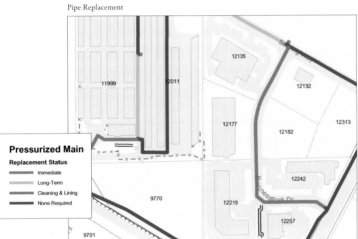

Pipe Replacement

Pressurized Main

Replacement Status

- Immediate
- Long-Term
- Cleaning & Lining
- None Required

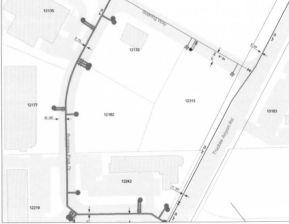

Valve Isolation Analysis

Truckee Donner Public Utility District

Truckee, California, USA

By Ian Fitzgerald

Contact

Ian Fitzgerald

ianfitzgerald@tdpud.org

Software

ArcFM, ArcMap 8.3, and custom applications

Hardware

Dell workstation

Printer

HP Designjet 1055cm

Data Source(s)

Data built internally

Utilities—Water/Wastewater

The Truckee Donner Public Utility District services the water needs of 15,000 customers over a 44-square-mile area. Previously, water facility information was maintained on old, decomposing paper and Mylar engineering documents, some of which are nearly 75 years old. Now, a seamless two-foot accurate water facility map and a fully functional three-dimensional system model have taken mapping to a new level. Improved facility maintenance, faster and more accurate system design, and more precise facility location are many functions available to users of the water GIS model.

Valve Isolation—Due to either construction or pipe leaks, pipe sections sometimes need to be temporarily taken out of service. This modeling capability enables the user to select the leak or construction point of the water network and perform a valve isolation trace. The analysis results will show, either by selection or visually, the critical valves that require closing and all water facilities and customers to be affected by the isolation.

Trench Management—Improved underground facility management is accomplished by Miner & Miner's Conduit System Manager's ability to model trenches. The conduit manager depicts pipes and/or conduits within a trench and their locations, and it color codes the utility using the conduit.

Pipe Replacement—To minimize leaks, improve the integrity of the water system, and efficiently manage fieldwork time, the pipe replacement application quickly and efficiently assigns each pipe segment a replacement designation. Based on a coefficient that considers pipe material, year installed, and number of leaks, each pipe is assigned a replacement cost and is later used to develop a replacement report for that month.

Three-Dimensional Plan and Profile—Water utility engineering is aided through the modeling of the water infrastructure in a three-dimensional view. The user has the ability to present a cross sectional view of any pipe segment in relation to street surface, gas pipes, and electric and broadband conduits. This is important to show true pipe location.

LEGEND

Distribution Poles
Transmission Poles
Lift Station
Substations
Transmission Lines
Buildings
Roads
Water
BPUB Buildings

ELEVATION
measured in feet
above mean sea level

> 40'
Between 37' and 40'
High : 37.000000

Low : 15.000000
Between 13' and 15'
Between 10' and 13'
Lower than 10'

100 Year Floodplain
500 Year Floodplain

This map was developed as a communication and information tool to be used by Brownsville Public Utilities Board (BPUB) management in the event of a hurricane. The map is updated and distributed each June. Digital elevation data is color coded and ranges from 10 to 40 feet in elevation. This data was emplaced along with utility facilities, such as electric poles and lift stations, and planimetric data from the BPUB GIS. Federal Emergency Management Agency flood zones were overlaid. This map highlights the areas and facilities that are of particular concern in a heavy rain event such as a hurricane.

BPUB, the water, sewer, and electric service provider for Brownsville, Texas, has developed a full-featured GIS based on ESRI technologies. BPUB is currently integrating applications that range from work order management systems to engineering design systems. By harnessing the power of GIS combined with customer information and accounting systems, each department is enabled with the tools to increase effectiveness and productivity.

Brownsville Public Utilities Board

Brownsville, Texas, USA
By David Bartle

Contact
Michael Elam
melam@brownsville-pub.com

Software
ArcMap 8.3, ArcSDE, Oracle, and Windows 2000

Hardware
Dell PC and Dell server

Printer
HP Designjet 5000

Data Source(s)
BPUB; Center for Space Research, University of Texas, Austin; Federal Emergency Management Agency

Utilities—Water/Wastewater

Sewer Cleaning Frequency Map

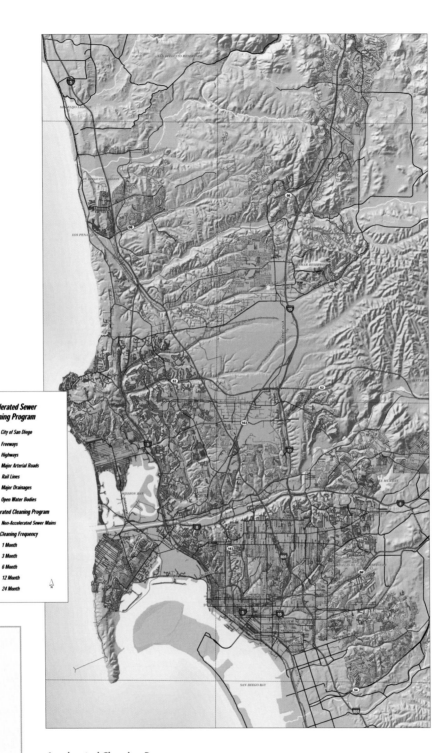

**City of San Diego—
Metropolitan Wastewater
Department**

San Diego, California, USA

By John Evans

Contact

Lisa Canning

lcanning@sandiego.gov

Software

ArcMap 8.3 and Windows 2000

Printer

HP Designjet 1055cm

Data Source(s)

City of San Diego, SanGIS, and
U.S. Geological Survey

Utilities—Water/Wastewater

Accelerated Cleaning Program

San Diego performs cleaning activities on all of its sewer
pipelines. Pipes with documented problems are cleaned more
frequently with some pipes getting cleaned as often as once
a month. Cleaning crews gather data in the field so the cleaning
frequencies can be modified as necessary. The map shows the pipes
on the accelerated cleaning program, which are the pipes that have
historically required the most maintenance.

Sewer Mains by Date Installed

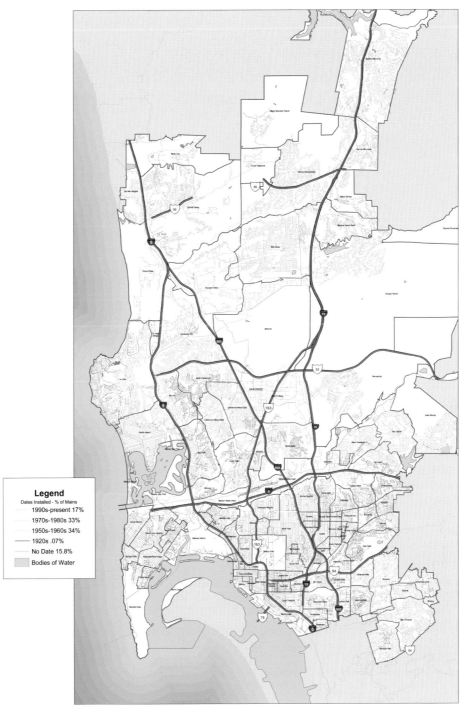

Legend

Dates Installed - % of Mains

— 1990s-present 17%
— 1970s-1980s 33%
— 1950s-1960s 34%
— 1920s .07%
— No Date 15.8%
Bodies of Water

San Diego's wastewater collection system consists of approximately 2,894 miles of sewer pipe made of various materials.

Noting the year of installation helps the city identify potential problem areas and address those issues in a systematic manner. The progress in the replacement of older pipes in each neighborhood can be seen graphically in this map as well as what remains to be done.

City of San Diego—Metropolitan Wastewater Department

San Diego, California, USA
By Lisa Canning and Ray Fischer

Contact

Lisa Canning
lcanning@sandiego.gov

Software

ArcMap 8.1 and Windows 2000

Printer

HP Designjet 750c

Data Source(s)

City of San Diego—Metropolitan Wastewater Department and SanGIS

Utilities—Water/Wastewater

Index by Organization

AAA, 99

ADAUHR, 12

Africa Institute of South Africa, 93

Agriculture Research Council—Institute for Soil, Climate, and Water, 66

AlburyCity (Local Government Authority), 51

Alpart Mining Venture, 78

City of Berkeley, 44

Bowne Management Systems, Inc., 103

Town of Brookline, 54

Brown University, 28

Brownsville Public Utilities Board, 117

C–K Associates Geospatial Technologies, 91

California Department of Transportation, 107

Cascade Environmental Resource Group Ltd., 36

Clarion University, 20

Clover Point Cartographics Ltd., 14

Coeur d'Alene Indian Tribe, 101

College of the Atlantic, 94

Conservation International, 18, 19

Converium Ltd., 4

CoServ Electric, 113

Country Fire Authority of Victoria, 43

CSIR Environmentek, 34

Czech Geological Survey, 80

Danish Institute for Fisheries Research, 86

Dartmouth College, 62

Dominion, 114

Ecology and Environment, Inc., 37

ECOWISE Environmental, 24

EDAW, 29

ENTRIX, Inc., 82

Environment Bay of Plenty, 87

Environmental Protection National Agency, 88

ESRI, 11, 81

Explorer Graphics Ltd., 85

Foothills Model Forest, 21

Forest Resources Institute, Arthur Temple College of Forestry, Stephen F. Austin State University, 6

Geodigital Ltda., 98

Geographic Data Technology, Inc., 17

Global Infosci, 60

Gwinnett County, 56

Haifa University GIS and Remote Sensing Lab, 25

Hillsborough County, 59

International Civil Aviation Organization Air Navigation Bureau AIS/MAP Section, 110

Indiana University, 35

Japan National Institute for Environmental Studies, 89

Liberec Region, 104

Los Alamos National Laboratory Earth and Environmental Science Division, 71, 72, 73

Meterologix, LLC, 81

Metropolitan Transportation Commission, 108

National Geospatial-Intelligence Agency, 61

National Oceanic and Atmospheric Administration, Pacific Marine Environmental Laboratory, 90

The National Power Company of Iceland, 112

The Norwegian National Rail Administration, 106

Parsons Corp., 109

Bureau of Planning, City of Portland, 50

PPL Electric Utilities, 111

Repsol YPF, 77

Riverside County Transportation and Land Management Agency GIS, 39, 55

City of Roswell, 47

San Diego Association of Governments, 42, 96, 97

City of San Diego—Metropolitan Wastewater Department, 118, 119

Sault Ste. Marie Innovation Centre, 84

Scientific and Projecting Institute of Spatial Planning ENKO, 26

South Florida Water Management District, 92

T-Kartor Sweden AB, 100

Texas Tech University, 22, 58

Tobin International, 81

City of Toronto, 53

Transport for London, 100

Truckee Donner Public Utility District, 115, 116

U.S. Air Force, Pacific Air Forces, 30

U.S. Army Corps of Engineers, Jacksonville District, 92

U.S. Bureau of Reclamation, 83

U.S. Department of Agriculture, Agricultural Research Service, National Soil Tilth Laboratory, 63

U.S. Department of Agriculture, National Agricultural Statistics Service, 64

U.S. Department of Agriculture, Natural Resources and Conservation Service, National Soil Survey Center, 63

U.S. Department of the Interior, Bureau of Land Management, 102

U.S. Environmental Protection Agency Pacific Southwest Region/Titan Corp., 33

U.S. Geological Survey, 16, 40, 41, 46, 75, 76

U.S. Geological Survey, Astrogeology Team, 13

U.S. Geological Survey, National Wetlands Research Center, 74

U.S. Geological Survey, Rocky Mountain Mapping Center, 8

U.S. Government, 38

United Nations Environment Programme, 70

University of California, Kearney Agricultural Center, 68

University of Houston—Geoscience Dept., 32

University of Illinois, 57

University of Utah, 10

University of Wisconsin–Eau Claire, 31

Verdi & Company, 5

Victoria Department of Sustainability and Environment, 43

Region Vysocina, 52

World Wildlife Fund Canada, 23

Weyerhaeuser Company Ltd., 69

City of Yakima, 48